Palgrave Studies in Climate Resilient Societies

Series Editor
Robert C. Brears
Avonhead, Canterbury, New Zealand

The Palgrave Studies in Climate Resilient Societies series provides readers with an understanding of what the terms **resilience and climate resilient** societies mean; the best practices and lessons learnt from various governments, in both non-OECD and OECD countries, implementing climate resilience policies (in other words what is 'desirable' or 'undesirable' when building climate resilient societies); an understanding of what a resilient society potentially looks like; knowledge of when resilience building requires slow transitions or rapid transformations; and knowledge on how governments can create coherent, forward-looking and flexible policy innovations to build climate resilient societies that: support the conservation of ecosystems; promote the sustainable use of natural resources; encourage sustainable practices and management systems; develop resilient and inclusive communities; ensure economic growth; and protect health and livelihoods from climatic extremes.

More information about this series at
http://www.palgrave.com/gp/series/15853

Rutger de Graaf-van Dinther
Editor

Climate Resilient Urban Areas

Governance, Design and Development in Coastal
Delta Cities

Editor
Rutger de Graaf-van Dinther
Rotterdam University of Applied Sciences
Rotterdam, The Netherlands

ISSN 2523-8124　　　　　　　ISSN 2523-8132　(electronic)
Palgrave Studies in Climate Resilient Societies
ISBN 978-3-030-57536-6　　　ISBN 978-3-030-57537-3　(eBook)
https://doi.org/10.1007/978-3-030-57537-3

This Palgrave Macmillan imprint is published by the registered company Springer Nature Switzerland AG.
The registered company address is: Gewerbestrasse 11, 6330 Cham, Switzerland

CONTENTS

List of Figures

LIST OF TABLES

CHAPTER 1

The Five Pillars of Climate Resilience

Rutger de Graaf-van Dinther and Henk Ovink

Abstract The urgency of climate resilience for urban areas is discussed by providing an overview of the most important climate impacts such as floods, droughts, biodiversity decline, urban heat stress, and social vulnerability. This book presents the overarching framework of climate resilience in urban areas. This framework consists of five pillars: threshold capacity, coping capacity, recovery capacity, adaptive capacity, and transformative capacity. This framework is based on an extensive literature review of definitions on resilience and vulnerability, and it can be used to develop comprehensive resilience strategies. The framework will be used as an integrating thematic line throughout all the chapters in the book.

R. de Graaf-van Dinther (✉)
Research Centre Sustainable Port Cities, Rotterdam University of Applied Sciences, Rotterdam, The Netherlands

Blue21 and Indymo, Delft, The Netherlands
e-mail: r.e.de.graaf@hr.nl

H. Ovink
Special Envoy for International Water Affairs at Kingdom of the Netherlands, The Hague, The Netherlands

Keywords Vulnerability • Resilience • Threshold capacity • Coping capacity • Recovery capacity • Adaptive capacity • Transformative capacity

1.1   Background

Climate change, population growth, and urbanisation challenge liveability and safety in vulnerable coastal cities. The importance of mitigating climate change is demonstrated by the 2015 Paris Climate Agreement in which most of the world's nations have agreed to take measures to keep the world well below an average temperature rise of 2 °C. At the same time, adaptation to climate change is crucial, as the impact of climate change is already becoming increasingly critical, in particular in densely populated, low-lying coastal areas.

1.1.1   Climate Impacts in Urban Areas

Cities in particular are vulnerable to climate change impacts (Rosenzweig et al. 2011). With more than half of the global population currently living in cities, climate change already has a profound impact on society. Climate change is increasing demonstrated in the number of extreme weather events across the world, most profoundly impacting urban regions. In the past decades, the number of extreme precipitation events and heatwaves has increased significantly worldwide (Mishra et al. 2015), again impacting our cities severely. Furthermore, the *urban heat island effect*, which increases urban temperature compared to surrounding areas, creates more temperature increases in addition to the effects of climate change (Huang et al. 2019). The negative impacts of excessive heat include a significant reduction in air quality, increased demand for water and energy, and a wide range of public health impacts, in particular among vulnerable social groups (Schar and Jendritzky 2004).

Climate change also severely impacts water quality and water availability in (growing) urban areas. Almost half of the world's 482 largest cities will face water shortages by 2050 (Flörke et al. 2018). They will experience a direct shortage of water and increased competition for scarce water resources with other water-demanding sectors such as agriculture. Without proper mitigation and adaptation measures (IFPRI 2012), approximately 45% of global GDP and 52% of the global population will be exposed to severe water scarcity by 2050.

Population growth is particularly strong in urban coastal areas and around rivers, in deltas that are vulnerable to sea level rise, floods, and other climate change impacts. It is estimated that the average global flood losses in coastal cities will multiply from US$6 billion per year in 2005 to US$52 billion per year by 2050, taking into account just the socio-economic costs of floods, such as the impact on populations and property values (Hallegatte et al. 2013). The total size of global urban areas exposed to both floods and droughts is expected to increase by more than 250% in 2030 compared to 2000 (Güneralp et al. 2015). In 2060, the global population in low elevation coastal zones could more than double compared to the year 2000 (Neumann et al. 2015) (Fig. 1.1).

Fig. 1.1 Flood impact affecting daily life in Semarang, Indonesia. ((c) Cynthia van Elk | Water as Leverage)

1.1.2 Decline in Biodiversity

In addition to climate impacts, the global decline in biodiversity is weakening the natural system that protects our economic and social systems (Cardinale et al. 2012). High-quality ecosystems contribute to people's health and quality of life. They purify and preserve water, store carbon, mitigate flooding, and deliver foods and fibres. Recent research (Ligtvoet et al. 2018) commissioned by the Netherlands Environment Agency (PBL) shows that the quality and functioning of aquatic ecosystems are influenced by developments in river basins and deltas: urbanisation, agricultural production, water abstraction for irrigation, water for industrial and domestic use, dams constructed for hydropower, and water pollution caused by emissions from agriculture, households, and industries. Apart from biodiversity loss, ecosystem services may also deteriorate. Examples of such services include clean drinking water, irrigation water, fish resources, carbon storage, recreation, and natural flood protection. The negative aspects of these developments particularly affect people who depend on natural resources for their livelihood.

Wetlands form part of some of the most productive and biodiverse habitats and are of great importance for the functioning of aquatic ecosystems. About 80% of the world's wetlands, both inland and coastal, have disappeared since 1700, and this loss is still continuing. The main causes are reclamation of wetlands for agriculture and urban development, canalisation for shipping, and construction of dams that reduce the inundation of wetlands and affect natural river flows. Wetland loss means also loss of carbon storage. Under the business-as-usual scenario, developments will result in further biodiversity loss in nearly 40% of the world's freshwater ecosystems (Ligtvoet et al. 2018).

The increase in nutrient emissions towards 2050 will result in an increase in nutrient loading to coastal waters, especially in the Asian region. This will increase the risk of toxic algal blooms and oxygen depletion in those waters, and will negatively affect biodiversity (e.g. coral reefs) and ecosystem functions, such as aqua-culture and fisheries (Cochard 2017).

1.1.3 Social Vulnerability

Informal settlements are the most exposed to climate change impacts, in many cities, especially in developing countries where inhabitants of such settlements make up more than 50% of the urban population. For example, in the Asian megacity Mumbai, which consists largely of informal settlements, 60% of its territory will be located in flood-prone areas by 2050 (Kulp and Strauss 2019). Water- and climate-related disasters disproportionately affect people living in such settlements (UNISDR 2009). Moreover, the resilience to bounce back after disasters is generally lower in these areas (Doberstein and Stager 2013) (Fig. 1.2).

Indigenous knowledge can contribute to reducing social vulnerability through the understandings, skills, and philosophies developed by societies with long histories of interaction with their natural surroundings. For many indigenous peoples, indigenous knowledge informs decision making about fundamental aspects of life, from day-to-day activities to longer-term actions. This knowledge is integral to cultural complexes, which also encompass language, classification systems, resource use practices, social interactions, values, rituals, and spirituality. These distinctive ways of knowing are important facets of the world's cultural diversity (UNESCO 2020).

Fig. 1.2 Climate-related disasters disproportionally affect people living in informal settlements. ((c) Cynthia van Elk | Water as Leverage)

1.1.4   How Cities Respond

Given the current and expected climate impacts in urban areas, cities have started initiatives to respond to these threats. Significant advances have been made in the fields of urban climate adaptation policy, research, and practice (Carter et al. 2015). Global networks for capacity building and knowledge exchange have emerged, such as ICLEI Local Governments for Sustainability (Acuto 2016), the C40 Cities Climate Leadership Group, United Cities and Local Governments (UCLG), and The Rockefeller Foundation's 100 Resilience Cities Initiative, which has now evolved into the Global Resilient Cities Network and the NGO Resilient Cities Catalyst.

Increasingly, the concepts of resilience and vulnerability are associated with cities' efforts to respond to, and prepare for, climate change. The concept of resilience emerged initially in ecology, where resilience was understood as the ability of ecological systems to persist in the face of disturbance and maintain functional system integrity (Holling 1973). In the last decades, the concept of resilience has been applied in a wide range of disciplines including economics, psychology, social sciences, natural hazards, and engineering (Bahadur et al. 2011). There are many different interpretations and definitions of both these concepts—vulnerability and resilience.

De Graaf, van de Giesen, and van de Ven (2009b) investigated literature definitions of vulnerability and resilience and derived from this a practical approach for the field of water management. This framework for resilience consists of four pillars: threshold capacity, coping capacity, recovery capacity, and adaptive capacity. For resilient water management, it is important to prepare and prevent damage resulting from environmental variation, to reduce damage during extreme weather events, to recover effectively after disasters, and to adapt to current and expected trends in the environment. More recently, a fifth component, transformative capacity, has been suggested (Ovink 2019). The impact of climate change on urban areas can be so sizeable that adaptation is no longer sufficient (Kates et al. 2012). Instead, proactive transformation of the urban system becomes a necessity. Olsson et al. (2010) defined the term, *transformative capacity*, as the capacity to transform socio-ecological systems' trajectories towards ecosystem stewardship. The practices and processes to build transformative capacity are "deeply participatory and co-creative" (Ziervogel et al. 2016, p. 7). Transformative capacity, however, not only includes the social system (formal and informal), but is also about the physical system (De Graaf and Van der Brugge 2010). Through identifying and implementing catalysing interventions, our physical systems can leapfrog towards a progressive state of resilience. These interventions are hotspots of the systems' vulnerability, and their capacity is to help leapfrogging towards resilience, among other things because of their capacity to be (easily) scaled up and replicated for progressive resilience impact. In this book, transformative capacity is added as a fifth component.

1.2   The Five Pillars of Climate Resilience

Climate resilience is defined in this book as consisting of five capacities or pillars: threshold capacity, coping capacity, recovery capacity, adaptive capacity, and transformative capacity.

1. Threshold capacity: the capability to prevent damage by constructing a threshold against environmental variation.
2. Coping capacity: the capability of a neighbourhood, city, or country to deal with extreme weather conditions and reduce damage during such conditions.
3. Recovery capacity: society's capability to bounce back to a state equal to, or even better than, before the extreme event.
4. Adaptive capacity: society's capability to anticipate uncertain future developments.
5. Transformative capacity: the capability to create an enabling environment, strengthen stakeholder capacities, and identify and implement catalysing interventions to transition proactively to a climate-resilient society.

This framework of five pillars of urban climate resilience is used as an integrative concept throughout the chapters, in which it will be evaluated whether the five capacities can be recognised in the different case studies presented, whether this leads to additional insights, whether it is necessary to address all five components to develop a comprehensive resilience strategy, and what the limitations of the framework are. Table 1.1 presents an overview of measures and approaches in water management to strengthen the five capacities.

Table 1.1 Examples of water management measures to strengthen the five capacities of urban climate resilience

Threshold capacity

Dikes	Water reservoirs
Flood barriers	Water supply networks

Coping capacity

Emergency plans	Temporary flood barriers
Forecasting and timely warning of extreme weather events	Flood-proof building materials
	Increasing risk awareness
Evacuation plans	Providing emergency shelter
First aid capacity	Neighbourhood assistance networks
Backup water supply and other utilities	Connections with other river basins
Flood-proof critical infrastructure	

Recovery capacity

Recovery plans	Disaster funds
Rehabilitation schemes	Flood-relief organisations
Equipment and spare parts available	Coordinated plans for interregional and international support
Insurance schemes	

Adaptive capacity

Developing innovation niches and transition experiments	Developing a portfolio of water resources including local sources
Experimenting with flood-proof modes of urbanisation, e.g. floating urban development	Building with nature
	Room for the river
Flexible and reversible flood-control infrastructure	Integrating flood management and spatial planning

Transformative capacity

Proactive inclusive planning and design with all stakeholders	Build potential for replication, improvement, upscaling, and mainstreaming of innovation projects
Strengthen local capacity for planning, execution, and maintenance	From single projects to consistent continuous innovation processes
Invest in enabling environment for comprehensive climate resilience strategy and implementation	Strategically link water issues to urban dynamics
Transdisciplinary process, dissolving sectoral, and disciplinary boundaries	Look for synergies between urban development and ecological restoration
Create coalitions of local and global stakeholders in inclusive learning environment	Leapfrogging phases in the development of water cities
	Regenerative urban developments
Build trust and safe environments, soft spaces	Develop the narrative to aspire and inspire and drive decisions/decision makers

1.2.1   Threshold Capacity

Threshold capacity is a society's ability to prepare and build up a threshold against variation in order to prevent damage. In flood risk management, examples include building river dikes and increasing flow capacity to set a threshold against high river flows. For water supply, examples include constructing storage reservoirs to increase the damage threshold by preventing loss of service in the event of droughts. The objective of building threshold capacity is damage prevention. In addition to physical infrastructures, nature-based solutions—our ecosystems—are very well equipped with this threshold capacity but are in rapid decline. Investing in nature-based solutions helps to increase our resilience with multiple benefits—reducing heat islands, increasing the health and the quality of our social, environmental, and economic systems and places. The time horizon of threshold capacity originates in the past; a society's past disaster experiences are the guiding principle determining the height of the threshold. In the Netherlands, for ages, dikes of the same height as the highest experienced flood were constructed; and the dimensions of a water resources reservoir are determined by historic droughts and water use levels. As a result, uncertainty about the height of the threshold is relatively low. A society's ability to build, operate, and maintain threshold capacity is determined by that society's environmental resources and its social, institutional, technical, and economic abilities. In the Netherlands, this is relatively well organised. The responsibility for the maintenance of flood-defence and water-delivery infrastructure is clear. Waterboards are responsible for maintaining flood defences; water utility companies are responsible for safe and efficient delivery of drinking water. Many other countries, however, lack the technical, institutional, and financial capacities to build a high threshold capacity.

1.2.2   Coping Capacity

Coping capacity is a society's capacity to reduce damage if a disturbance exceeds the damage threshold. For flood management, a society's coping capacity is determined by the presence of effective emergency and evacuation plans, the availability of damage reducing measures—again nature-based solutions, a communication plan to create risk awareness among inhabitants, and a clear organisational structure and responsibility for disaster management. Early warning systems are also part of the wide variety of instruments that a society has at its disposal. With the necessary ownership embedded in social and operational structures, these mechanisms help increase a society's coping capacity. For water supply, the availability of emergency and backup water facilities that can be used in the event of droughts and disasters are important determinants of coping capacity. The objective of developing coping capacity is damage reduction, either by reducing flood impacts or by reducing loss of the water supply service. The time orientation is instantaneous, because in emergencies, only the here-and-now is important. The uncertainty is low because the magnitude of the hazard is clear at the time that a society has to deal with it. Furthermore, the ability of a society to build, operate, and maintain coping capacity is determined by that society's social, institutional, technical, and economic abilities. There is a large range of coping capacity options. In the Netherlands, threshold-exceeding events for water management do not occur frequently. This may explain why it is not clear who is responsible for damage reduction in emergencies. Multiple actors such as firefighters, waterboards, municipalities, and other government agencies are involved.

1.2.3   Recovery Capacity

Recovery capacity is the third component and refers to a society's capacity to recover to a state that is the same as, equivalent to, or better ("build back better") than, before the emergency. For flood control, it is the capacity in a flooded area to reconstruct buildings, infrastructure, and dikes. For water supply, it is the capacity to re-establish a functioning water supply and sanitation system. The objective of developing and increasing recovery capacity is to respond quickly and effectively after a disaster. The time horizon is instantaneous, right after the disaster, but will change gradually towards a focus on the future. Although economic damage estimates may be difficult, the uncertainty of the hazard magnitude will be relatively low because the effects will still be noticeable. The country's economic capacity to finance the reconstruction determines the recovery's success to a large extent. However, institutional ability and technical knowledge are also important. A society that is able to recover better (and fast) from impacts of hazards will be less vulnerable to these hazards. Incorporating future standards based on risks and vulnerability assessments is critically important from all perspectives: environmental, to safeguard further degradation and increase ecosystem resilience; social, for societies to invest in a future with capacity—enabling environment—and trust; economic, to prevent investments going to waste because of outdated standards and increased future risks, and, because the value of these investments has a societal return, this increases economic resilience. Recovery time may range from weeks to decades, depending on the spatial scale and disaster magnitude. Recovery from Hurricane Katrina in New Orleans was projected to take years (Kates et al. 2006). Although in the Netherlands it is clear who is responsible for reinstalling the flood control and water delivery infrastructure, it is not entirely clear who is financially responsible for compensating the hazard impacts. The Dutch government often refunded flood damage to houseowners in the past. However, people themselves or insurance companies could also be responsible, in particular in areas that are not protected by dikes.

1.2.4   Adaptive Capacity

Adaptive capacity is a society's capacity to anticipate uncertain future developments. This includes catastrophic, infrequently occurring disturbances like extreme floods and severe droughts. The time orientation of adaptive capacity lies in the future. Although a system may be functioning well at present, human and environmental developments, from both inside and outside the considered system, can put a system under strain and threaten its future functioning. Examples include climate change, population growth, and urbanisation. The acknowledgement that these processes may be influenced but cannot be predicted, engineered, or controlled is central to the importance of adaptive capacity. Because the system cannot be optimised for a known situation in the future, it is important to build adaptive capacity by anticipating uncertainty. Another reason to develop adaptive capacity is the acceptance that dealing with these uncertain future developments might require more than improving threshold, coping, and recovery capacity. From this perspective, developing adaptive capacity is a form of the precautionary principle. Without adaptive capacity, a society will try to recover from climate change impacts until it is no longer possible.

Preventing a technical lock-in pattern and securing diversity by keeping options open for future development contribute to adaptive capacity (Folke et al. 2002; Pahl-Wostl 2007). New technologies and innovations will be developed in the future. Adaptive infrastructure means that these options can be incorporated in large technical water management systems.

Without adaptive capacity, promising new technologies that are not compatible with the current system will be excluded, and opportunities will be missed. Water management systems should therefore be flexible and reversible to allow future changes to be made. Therefore, adaptive capacity improves the freedom of future generations to implement alternative options. The role of experts is to make as many options available to society as possible and place these options in relation to society's objectives (Harremoës 1997). Technologies reflect societal values; they are socially constructed (e.g. Bijker 2006). Adaptive capacity is therefore also a necessity for ethical reasons. Adaptive capacity offers freedom to future generations because it enables them to include technologies, ways of operation, governance, values, and culture in water management infrastructure that reflect their values.

For flood control, the problem of adapting to uncertain future developments can be illustrated by an example of land use. Although future risks from river or sea floods are unknown, land use decisions that determine future vulnerability are currently being taken. For water supply, a good example is salt water intrusion. The sea level and river discharge in 2050 can only be sketched in scenarios with wide ranges of implications, but with clear interdependencies across the vulnerabilities and opportunities in and across all systems; hence, the future problem of salt water intrusion in river deltas is also more unpredictable. However, the consequences of decisions to construct inlets for drinking water production points in these deltas exceed the horizon of reliable climate predictions.

The objective of developing adaptive capacity is to anticipate future developments and impacts by constructing a robust living and working environment. The uncertainty about the nature and magnitude of future hazards and impacts is high and the frequency of occurrence is low, although increasing. The capacity to adapt to these uncertain developments also determines a system's vulnerability. Although the exact size and nature of changes are unknown, solutions will have to be developed for long-time horizons, and financial and spatial reservations will have to be made to allow for adaptations. The IPCC (2001) presented many options available for society to increase its adaptive capacity, varying from technical options to insurance policies and communication strategies. The range and variety of possible adaptive options is large, and the number of organisations involved in the adaptive capacity determinants is also large. Consequently, in most countries, there is no clear picture about who is responsible for strengthening adaptive capacity. In the

Netherlands, recently the Delta Programme has incorporated adaptive delta management, which was approved by Parliament, and therefore adaptation now entails more political responsibility (Delta Programme Commissioner 2020).

1.2.5   Transformative Capacity

Transformative capacity is a society's capacity to transform itself in face of expected catastrophic developments such as human-induced climate change impacts. Similar to adaptive capacity, the time orientation of transformative capacity lies in the future. The main difference is that adaptation is more associated with small step incremental change in the current system, whereas transformation is regarded as transforming the current system into a system with fundamentally different system characteristics (Kates et al. 2012). Innovation plays an important role in both adaptation and transformation processes, as they require new stakeholder roles, new spatial processes, new guidelines, new user practices, and new knowledge. However, adaptive innovations are often implemented as part of the current system and are mostly applied on a relatively small scale, for instance in pilot projects. Transformative innovation aims to proactively transform the entire system or a societal sector, such as the water management sector. Evaluating, improving, and learning from pilot projects to make them suitable for upscaling and mainstreaming to a larger scale are important components of transformative capacity.

Stakeholder involvement and receptivity are key concepts for investigating the conditions required for transformative change and thus a crucial component of transformative capacity (De Graaf et al. 2009a). Jeffrey and Seaton (2003, pp. 281–282) defined receptivity as "the extent to which there exists not only a willingness (or disposition) but also an ability (or capability) in different constituencies (individuals, communities, organisations, agencies, etc.) to absorb, accept and utilise innovation options." Stakeholder receptivity consists of awareness, association, acquisition, and application. Awareness is the perceived sense of urgency among stakeholders that the current system, even when adapted, is not capable of dealing with the expected threats and that a system transformation is necessary. System understanding is a more advanced level of awareness. Understanding is the capacity to have insight into all vulnerabilities, interdependencies, and opportunities and to relate this to values and ways to operate and manage the water system (Laeni et al. 2020, this book). Association is the ability of stakeholders to link transformative change to their own agenda, objectives, and interests. Acquisition is the ability to acquire, implement, operate, and maintain the alternative system. Application is the ability to have sufficient legal and financial incentives to apply the alternative options.

In addition to stakeholders' capacities, an enabling context or enabling environment should be present, developed, or strengthened for transformative change. Brown and Clarke (2007) introduced an interplay between champions and change agents on the one hand and an enabling context on the other hand as a required mechanism for the transition to sustainable urban water management. Components of enabling capacity include social capital, trusted and reliable science, strategic funding, and market receptivity. Pahl-Wostl (2017) outlined the governance properties of transformative capacities. The transformation of socio-technical systems involves changing multiple interlinked systems such as water management, energy, transportation, and land use planning, at different scales. For this purpose, water should be strategically linked to other societal objectives and urban dynamics. Moreover, stakeholders should be facilitated to cooperate in a transdisciplinary process, dissolving sectoral and disciplinary boundaries. There should be effective links between the informal setting and formal policy processes (De Graaf and Van der Brugge 2010; Werbeloff and Brown 2016). Next, there should be a balance between top-down and bottom-up modes of governance. Sustaining transformational change over time requires a continuous process and a supportive social context in which the availability of resources for action and acceptable and understandable options are key enabling factors (Kates et al. 2012) (Fig. 1.3).

There is a clear need for time and space to be able to achieve a really inclusive process, based on trust, without power dependencies and mixed interests getting in the way. Process and place can be determined as soft spaces (Allmendinger and Haughton 2009). These are areas where deliberate attempts are made to introduce new and innovative ways of thinking, especially in places where there is considerable resistance to cross-sectoral and inter-territorial governance. Design thinking, acting, and the design process are crucial for the transformative approach and its impact. Three pillars of design capacity were outlined by Ovink and Boeijenga (2018): (1) design is solution oriented aimed at innovative, catalytic, and pragmatic solutions to known and unknown challenges—getting things done, where it matters most, now and in the future; (2) design is intrinsically holistic, it has the capacity to connect needs and opportunities across scale,

Fig. 1.3 Transformative capacity requires an inclusive process based on trust. ((c) Cynthia van Elk | Water as Leverage)

time, and interests; and (3) design is political, with its narrative capacity and its aspirational and inspirational character.

1.3   Scope of the Book

Many cities have taken up the urgent challenge to become resilient to the impacts of climate change. This challenge goes further than the ability to resist the impacts of extreme conditions. Coping with climate impacts and the ability to recover from them are equally important, as well as the capacity to adapt to the effects of climate change and the ability to transform the urban system. Resilience is therefore an intrinsically inclusive and holistic concept that includes various themes, for instance technical resilience against natural hazards such as floods, droughts, and other extreme weather events. It also has strong social and governance dimensions, such as the presence and strength of neighbourhood social assistance networks in the event of disasters and stakeholders' capacity to innovate their working practice, enabling the transformation of their city. For coastal cities, the challenge of the resilience journey means utilising scientific knowledge, but also the knowledge of citizens, indigenous peoples, and practitioners. Measures and strategies on different scales are needed from the national scale all the way down to neighbourhood, street, and building level. In some cases, optimising the existing urban infrastructure might be sufficient. More often, a transformation of the urban governance system is needed: resilience is by its essence systemic, and resiliency interventions are also systemic. The systems level of impact often is not met with a suitable governance (formal and informal) system. To develop transformative interventions, this means creating soft spaces for transformation. For implementation and operations, this implies system changes in governance, collaborative models, and coalitions. The potential of innovative pilots to improve, replicate, and scale up is a key factor for transformative change. This book explores the holistic nature of climate resilience in urban areas. It includes insights on different scales from areas of expertise such as engineering, social sciences, and urban design. Besides scientists from different fields, leading practitioners working in various global coastal cities have contributed to the book. Many of the book contributions are from the Netherlands, a country that is already partly located below sea level and that has centuries of experience of living with the threat of water. However, the book also includes experiences and insights from coastal cities all over the world to present a global perspective.

References

Acuto, M. (2016). Give cities a seat at the top table. *Nature, 537*, 611–613. https://doi.org/10.1038/537611a.

Allmendinger, P., & Haughton, G. (2009). Soft spaces, fuzzy boundaries, and metagovernance: The new spatial planning in the Thames Gateway. *Environment and Planning A, 41*(3), 617–633.

Bahadur, A. V., Ibrahim, M., & Tanner, T. (2011). *The resilience renaissance? Unpacking of resilience for tackling climate change and disasters* (Strengthening Climate Resilience Discussion Paper 1). Brighton: Institute of Development Studies.

Bijker, W. E. (2006). *Of bicycles, bakelites, and bulbs: Toward a theory of socio-technical change.* Cambridge, MA: MIT Press.

Brown, R. R., & Clarke, J. M. (2007). *Transition to water sensitive urban design: The story of Melbourne, Australia* (Report No. 07/1), Facility for Advancing Water Biofiltration, Monash University, Melbourne, Australia.

Cardinale, B., Duffy, J., Gonzalez, A., et al. (2012). Biodiversity loss and its impact on humanity. *Nature, 486*, 59–67. https://doi.org/10.1038/nature11148.

Carter, J., Cavan, G., Connelly, A., Guy, S., Handley, J., & Kazmierczak, A. (2015). Climate change and the city: Building capacity for urban adaptation. *Progress in Planning, 95*, 1–66. https://doi.org/10.1016/j.progress.2013.08.001.

Cochard, R. (2017). Coastal water pollution and its potential mitigation by vegetated wetlands: An overview of issues in Southeast Asia. In G. Shivakoti, U. Pradhan, & H. Shiwakoti (Eds.), *Redefining diversity & dynamics of natural resources management in Asia. Volume 1, Sustainable natural resources management in dynamic Asia* (1st ed., pp. 189–230). Amsterdam: Elsevier.

De Graaf, R., & van der Brugge, R. (2010). Transforming water infrastructure by linking water management and urban renewal in Rotterdam. *Technological Forecasting and Social Change, 77*(8), 1282–1291.

De Graaf, R. E., Dahm, R. J., Icke, J., Goetgeluk, R. W., Jansen, S. J. T., & van de Ven, F. H. M. (2009a). Receptivity to transformative change in the Dutch urban water management sector. *Water Science and Technology, 60*(2), 311–320.

De Graaf, R., van de Giesen, N., & van de Ven, F. (2009b). Alternative water management options to reduce vulnerability for climate change in the Netherlands. *Natural Hazards, 51*(3), 407–422.

Delta Programme Commissioner. (2020). *Adaptive delta management.* https://english.deltacommissaris.nl/delta-programme/what-is-the-delta-programme/adaptive-deltamanagement. Visited 18 Feb 2020.

Doberstein, B., & Stager, H. (2013). Towards guidelines for post-disaster vulnerability reduction in informal settlements. *Disasters, 37*(2013), 28–47.

Flörke, M., Schneider, C., & McDonald, R. I. (2018). Water competition between cities and agriculture driven by climate change and urban growth. *Nature Sustainability, 1*, 51–58. https://doi.org/10.1038/s41893-017-0006-8.

Folke, C., Carpenter, S., Elmqvist, T., Gunderson, L., & Holling, C. (2002). Resilience and sustainable development: Building adaptive capacity in a world of transformations. *Ambio: A Journal of the Human Environment, 31*(5), 437–440.

Güneralp, B., Güneralp, İ., & Liu, Y. (2015). Changing global patterns of urban exposure to flood and drought hazards. *Global Environmental Change, 31*, 217–225.

Hallegatte, S., Green, C., Nicholls, R. J., & Corfee-Morlot, J. (2013). Future flood losses in major coastal cities. *Nature Climate Change, 3*(9), 802–806.

Harremoës, P. (1997). Integrated water and waste management. *Water Science and Technology, 35*(9), 11–20.

Holling, C. S. (1973). Resilience and stability of ecological systems. *Annual Review of Ecology and Systematics, 4*, 1–23.

Huang, X. L., Liu, X., & Seto, K. C. (2019). Projecting global urban land expansion and heat island intensification through 2050. *Environmental Research Letters, 14*(11). https://doi.org/10.1088/1748-9326/ab4b71.

IFPRI. (2012). *2012 Global Hunger Index, chapter 3: Sustainable food security under land, water, and energy stresses.* Washington, DC: International Food Policy Research Institute.

IPCC. (2001). *Impacts, adaptation, and vulnerability for climate change, third assessment report of the IPCC.* Cambridge, UK: Cambridge University Press.

Jeffrey, P., & Seaton, R. A. F. (2003). A conceptual model of 'receptivity' applied to the design and deployment of water policy mechanisms. *Environmental Sciences, 1*, 277–300.

Kates, R. W., Colten, C. E., Laska, S., & Leatherman, S. P. (2006). Reconstruction of New Orleans after Hurricane Katrina: A research perspective. *Proceedings of the National Academy of Sciences of the United States of America, 103*, 14653–14660.

Kates, R., Travis, W., & Wilbanks, T. (2012). Transformational adaptation when incremental adaptations to climate change are insufficient. *Proceedings of the National Academy of Sciences of the United States of America, 109*, 7156–7161.

Kulp, S. A., & Strauss, B. H. (2019). New elevation data triple estimates of global vulnerability to sea-level rise and coastal flooding. *Nature Communications, 10*, 4844. https://doi.org/10.1038/s41467-019-12808-z.

Laeni, N., Ovink, H., Busscher, T., Handayani, W., & van den Brink, M. (2020). A transformative process for urban climate resilience: The case of Water as Leverage Resilient Cities Asia in Semarang, Indonesia. In R. de Graaf-van Dinther (Ed.), *Climate resilient urban areas. Governance, design and development in coastal delta cities.* Palgrave Macmillan.

Ligtvoet, W., Bouwman, A., Knoop, J., de Bruin, S., Nabielek, K., Huitzing, H., Janse, J., van Minne, J., Gernaat, D., van Puijenbroek, P., de Ruiter, J., & Visser, H. (2018). *The geography of future water challenges. PBL infographics report.* The Hague: PBL Netherlands Environmental Assessment Agency.

Mishra, V., Ganguly, A. R., Nijssen, B., & Lettenmaier, D. P. (2015). Changes in observed climate extremes in global urban areas. *Environmental Research Letters, 10*. https://doi.org/10.1088/1748-9326/10/2/024005.

Neumann, B., Vafeidis, A. T., Zimmermann, J., & Nicholls, R. J. (2015). Future coastal population growth and exposure to sea-level rise and coastal flooding – A global assessment. *PLoS One, 10*, e0118571.

Olsson, P., Bodin, Ö., & Folke, C. (2010). Building transformative capacity for ecosystem stewardship in social-ecological systems. In R. Plummer & D. Armitage (Eds.), *Adaptive capacity and environmental governance* (pp. 263–286). New York: Springer.

Ovink, H. (2019). Personal communication. 22 Feb 2019.

Ovink, H., & Boeijenga, J. (2018). *Too big: Rebuild by design: A transformative approach to climate change*. Rotterdam: NAi010 Publishers.

Pahl-Wostl, C. (2007). Transitions towards adaptive management of water facing climate and global change. *Water Resources Management, 21,* 49–62.

Pahl-Wostl, C. (2017). An evolutionary perspective on water governance: From understanding to transformation. *Water Resources Management, 31,* 2917–2932.

Rosenzweig, C., Solecki, W. D., Hammer, S. A., & Mehrotra, S. (2011). *Climate change and cities: First assessment report of the urban climate change research network*. Cambridge, UK: Cambridge University Press.

Schar, C., & Jendritzky, G. (2004). Climate change: Hot news from summer 2003. *Nature, 432,* 559–560.

UNESCO. (2020). *Indigenous knowledge and climate change*. https://en.unesco.org/links/climatechange. Visited 18 Feb 2020.

UNISDR. (2009). *Global assessment report on disaster risk reduction 2009 – Risk and poverty in a changing climate*. United Nations International Strategy for Disaster Reduction (UNISDR), UNISDR Secretariat, Geneva, Switzerland, 207 pages.

Werbeloff, L., & Brown, R. R. (2016). Using policy and regulatory frameworks to facilitate water transitions. *Water Resources Management, 30*(11), 3653–3669. https://doi.org/10.1007/s11269-016-1379-6.

Ziervogel, G., Cowen, A., & Ziniades, J. (2016). Moving from adaptive to transformative capacity: Building foundations for inclusive, thriving, and regenerative urban settlements. *Sustainability, 8,* 955. https://doi.org/10.3390/su8090955.

CHAPTER 2

Integration of Water Management and Urban Design for Climate Resilient Cities

Nanco Dolman

Abstract Urban design can play a key role in addressing a wide range of climate-related water challenges such as water pollution, water scarcity, floods, land subsidence, storm water management, ecosystem services and public health. Both in urban retrofit projects as well as new urban development integration of water management in the different phases of design and development is important. Design and planning approaches such as water-sensitive urban design (WSUD) provide useful tools for strengthening the integration of water in spatial planning and urban design processes, requiring any spatial intervention or new development to be evaluated on opportunities for sustainability and innovation. WSUD focuses on the integration of the natural environment and sustainable technology in planning for urban water, combining hydrology, landscape architecture and sociology. The chapter concludes with recommendations on how cities can build on their own experience and lessons from practical cases to achieve more water-sensitive urban design.

N. Dolman (✉)
Royal HaskoningDHV, Amersfoort, Netherlands
e-mail: nanco.dolman@rhdhv.com

R. de Graaf-van Dinther (ed.), *Climate Resilient Urban Areas*,
Palgrave Studies in Climate Resilient Societies,
https://doi.org/10.1007/978-3-030-57537-3_2

21

Keywords Blue-Green Infrastructure • Climate Adaptation • Urban
Planning • Water Management • Water-Sensitive Urban Design

2.1   Introduction

The water that brings life and energy to our cities also poses a significant
threat. Countries often set up and maintain their key economic interests in
the areas that directly border coasts and rivers and inherently expose themselves to risks from water. The need to maintain life in deltas around the
world requires us to either dramatically invest in new flood protection
infrastructure or in adaptation.

Just as infrastructure and the choice of development site are key factors
in planning considerations, so water management has a legitimate claim to
be considered in this process. It is not just about open water and riverbanks: the choice of rainwater and wastewater systems also determines the
water storage facilities required or the water action plan.

Water is a connecting challenge in making cities resilient. The water
issue then becomes a way of looking at cities and at how to make them
climate- and future-proof (Dolman and Ogunyoye 2018).

There is a recognised need for a fundamental change in how cities manage water in response to increasingly frequent and extreme rainfall events,
drier summers and urban expansion. Approaches centred on "living with
and making space for water" are increasingly adopted internationally and
address the full water spectrum (floods to droughts). These include the
Dutch "Room for the River" programme and Australian water-sensitive
urban design (WSUD) initiatives.

Nature-based solutions (NBS), sustainable drainage systems (SuDS)
and Blue-Green Infrastructure (BGI, e.g. green roofs, swales, rain gardens, detention basins and ponds) are widely employed. These approaches
enrich society through the provision of multiple co-benefits, including
access to public greenspace, recreational opportunities, aesthetic enhance-
ments and improved management of environmental processes such as
flooding, drought, urban heat, water and air pollution.

This makes it possible to move towards a balanced approach to treat
precipitation close to where it lands through measures that deliver multiple benefits for water management alongside environmental and amenity
benefits. New and retrofit sustainable drainage systems are being developed in appropriate places and integrated with programmes to deliver

better and larger sewer systems. And taking a spatial planning approach in being flexible in building development, by building adaptively with water in mind and by investing in green infrastructure (Dolman et al. 2013). By strengthening green infrastructure and giving water more space in both the public and private domains, the water city of the future has the potential to grow into a blue-green city.

Achieving a blue-green and climate resilient city is above all about achieving a healthy and liveable city. Involving our living environment and social values is vital. After all, we are part of the urban ecosystem or "Ecopolis" (Tjallingii 1996).

2.2   Urban Water Transitions (UWT) Framework

Cities are experiencing the impacts of climate change through water-related disasters, while the sustainable management of water resources remains crucial for urban climate resilience. Accordingly, frameworks that integrate urban water management with climate change adaptation become increasingly relevant.

Like a spatial transition, urban development has a strong relationship with its water systems. Based on a historical analysis of the technical and institutional arrangement in urban water management over time, this is captured by the urban water transitions (UWT) framework (Brown et al. 2008), shown in Fig. 2.1.

The UWT framework identifies six distinct development stages that cities go through when they progress towards greater water sensitivity. A city's path towards greater water sensitivity has traditionally followed a sequential path whereby each "state" builds on the development of the previous stage.

The first three stages of the UWT framework describe the evolution of the water system to provide essential services such as secure access to potable water (Water Supply City), public health protection (Sewered City) and flood protection (Drained City). These are followed by the Waterways City, Water Cycle City and ultimately a Water-Sensitive City, which describe the anticipated evolution of the urban water system to deliver higher order services such as social amenity and environmental protection, provide reliable water services under constrained resources and ensure intergenerational equity and resilience to climate change.

In the water-sensitive city, the built and natural environments are in balance, and the sustainable use of rainwater, groundwater, surface water,

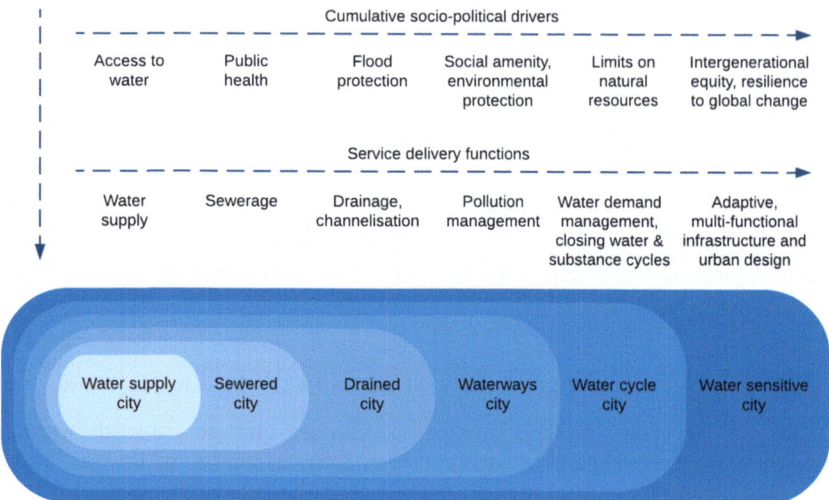

Fig. 2.1 Characteristics of the city-states in the Urban Water Transitions framework. (Hoekstra et al. 2018, adopted from Brown et al. 2008)

wastewater and drinking water is taken for granted. Ecosystems, infrastructure, communities, authorities and industry stakeholders collaborate to achieve resilience and empowerment for the future. When transitioning to a water-sensitive city, we face challenges in increasing experience, creating multiple use of space, combining functions and promoting education. The aim is to achieve practical and integrated solutions through improved awareness of water in design and planning, by improving processes and through the application of sustainable technology.

The six city-states are mapped against two dimensions:

1. **Cumulative Socio-Political Drivers**: "the socio-political drivers (demands and expectations) that emerge from society's growing environmental awareness, amenity expectations and evolving attitudes toward water management" (Brown et al. 2016).
2. **Service Delivery Functions**: "the increasingly diverse services required to address those drivers as cities transition to greater sustainability" (Brown et al. 2016).

To make the transition towards greater water sensitivity, cities have to consider three pillars of actions that integrate the governance, infrastructure and ecosystems dimensions of urban resilience under the following principles:

1. **Cities as water-sensitive communities and networks**: The implementation of integrated solutions requires improved perception of the benefits from decision makers, businesses and the public across multiple levels of governance. This makes collaboration a key requirement.
2. **Cities as water catchments**: The urban water system is often part of a larger catchment area. The intensive exploitation of the urban landscape can result in the progressive decline of the natural water system to the detriment of the surrounding region. The goal is to restore the water balance between these regions.
3. **Cities as ecosystem services providers**: Ecosystem services are the benefits that people derive from ecosystems. A river area for instance can be used multifunctionally for flood protection, groundwater recharge, recreation and the improvement of the quality of life. So, the water that poses a threat to society also brings life and energy to cities.

These three pillars of the Water-Sensitive City are strongly related or even based on the Ecopolis model (Tjallingii 1996) that links sociology to ecology by understanding:

1. The participating city—management of actors.
2. The living city—sustainable urban areas.
3. The responsible city—sustainable flow management (energy, water, waste, traffic, etc.).

More than 50% of cities aren't equipped with a sewer system or storm drainage, so developing cities have the potential to leapfrog towards greater water sensitivity through the provision of multifunctional and multipurpose water infrastructure. Yet, many developed cities have historically heavily invested in single-purpose systems, and consequently the maintenance and upkeep of these systems. The sustainable water usage and water-sensitive city transition for developed cities can be promoted in coherence with other transitions, for example, energy, circular economy.

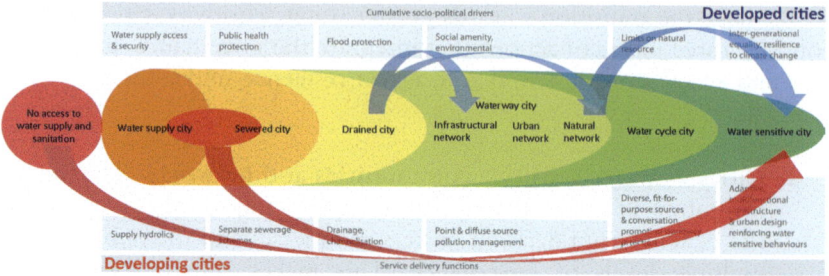

Fig. 2.2 Step-by-step approach of developed cities and potential of leapfrogging of developing cities in UWT framework. (Dolman and Ogunyoye 2019, adapted from Brown et al. 2008)

Currently, the EU-funded Interreg CATCH project is looking into the application of the Water-Sensitive City framework for climate adaptation in the North Sea Region (Dolman and Ogunyoye 2019). A step-by-step approach for developed cities has been proposed in three "Waterway City" substates (Fig. 2.2):

(a) Follows infrastructural network—adaptive,
(b) Co-organising in urban network,
(c) Retro-fit natural network—regenerative.

Taking this longer-term view, leapfrogging from "Waterway City" to "Water Sensitive City" could also be possible for developed cities.

2.3   Design and Planning Approaches

2.3.1   Multi-layer Safety as a Central Concept

All cities have lost rivers. Rivers were filled or put in culverts to contain the smell and to make flat land available. Buildings were constructed over the river itself, combined with raising the boggy land of the flood plain with waste materials.

Nowadays flood runoff is much greater even before we consider climate change, but the culverts' flood capacity has not increased. This can have serious consequences. An open river will have a higher flood capacity than a culvert, and a slight overflow won't have the catastrophic consequences of a blockage or collapse. "Daylighting" is the action of returning a culverted river to open water. At its simplest it is taking the lid off the culvert, but most designs aim to create a more natural river shape and re-introduce ecological habitats.

The fact that the climate is changing is evident from rising sea levels. Rivers will have to drain more water at certain times, and heavy rainfall that is difficult to absorb locally will occur more frequently, both resulting in flooding and damage.

The concept of multi-layer safety aims to make the water safety policy more robust and sustainable. A condition for this sustainability is that administrators, businesses and citizens are persuaded of the flood risk and take this into account in their decisions and how they act. Maintaining and strengthening water-aware behaviour require ongoing input.

Multi-layer Safety as a Central Concept (National Water Plan 2009)
The central concept of the water safety policy is "multi-layer safety":

1. The prevention of floods through strong dikes, dunes and flood defences (more robust and future-focused). Prevention remains the primary pillar of the policy.
2. Sustainable town and country planning with careful choice of location and consideration of land use issues to limit the numbers affected and amount of damage if flooding occurs.
3. Disaster management—good preparation is essential for effective response to a flood disaster, and will help to limit the amount of damage and the numbers affected.

2.3.2 Living with Water Principles

In cities worldwide, the processes of urbanisation and urban growth have altered the natural hydrological cycle by introducing impervious surfaces (roads, buildings, parking lots, etc.). In response, rainfall cannot infiltrate the ground and surface water runoff is much higher when compared to the natural state. A typical block of impervious surface is estimated to result in approximately five times more runoff than a natural area of the same size (USEPA 2003), depending on the local climate characteristics and soil type (Zevenbergen et al. 2012).

Traditionally, the approach to storm water management in the cities is through grey infrastructure of underground pipes designed to move water away from the city rapidly (Waggoner, Dolman et al. 2013). Many cities and countries are already recognising the need for broader and more

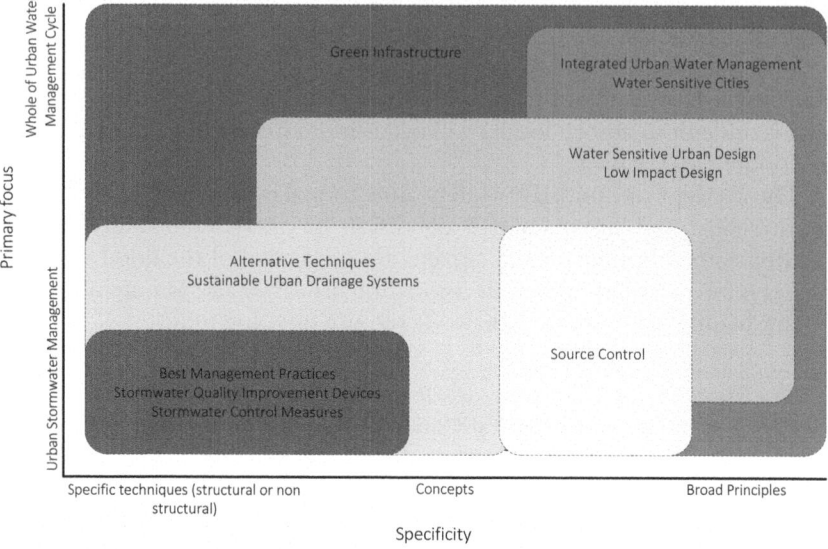

Fig. 2.3 Different terms for broader more sustainable approaches in urban water management. (Šakić Trogrlić et al. 2018, adopted from Fletcher et al. 2014)

sustainable approaches. As presented in Fig. 2.3, green infrastructure, best management practices (BMPs), low-impact development (LID), sustainable urban drainage systems (SuDS) and water-sensitive urban design (WSUD) are some of the names used for the various comparable measures and new paradigm in urban water management (Fletcher et al. 2014).

"Living with water" Principles
The pursuit of a safe, healthy and sustainable water management is of national interest in the Netherlands. Themes like "water and the city", "water as co-planning principle" and "sustainable and resilient water system" are identified as priorities in the Dutch national policy. Introduced in 4th Water Management Memorandum (1998) and included in 21st Century Water Management report (2000), this is concluded in the two three-stage strategies for:

- Water quantity (retain, store, drain).
- Water quality (prevent, separate, treat).

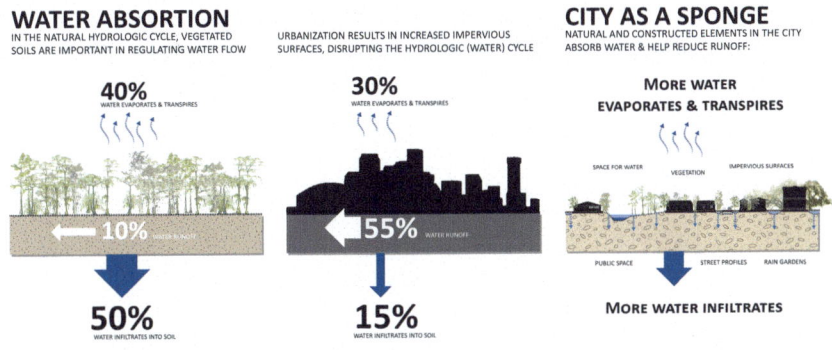

(source of % data: US Environmental Protection Agency, 1993)

Fig. 2.4 Imbalance of natural and urban water system and city as a sponge (Waggonner et al. 2014)

Focusing on water in city design and planning, the "living with water" principles to reduce vulnerability of residential areas in cities can be considered. Preferably these are translated into interventions or ecosystem services to balance the natural and urban water system (Fig. 2.4).

2.3.3   Spatial Adaptation as a Planning Strategy

Multi-layer safety as a central concept and the living with water principles can be integrated in spatial adaptation as a planning strategy. A first ever strategy of this kind was elaborated in the Resist, Delay, Store, Discharge (RDSD) strategy for the City of Hoboken (New Jersey, USA) as part of the Rebuild by Design competition in 2013 (Sect. 2.6).

Following the Delta Works (from 1953) and Room for the River (from 1997), spatial adaption has been adopted in the Netherlands as a whole of government approach. The Dutch Delta Plan on Spatial Adaptation was introduced in 2017 and is a collective plan, drawn up by municipalities, district water boards, provinces and the central government to render the Netherlands climate-proof and water resilient by 2050. The Delta Plan expedites and intensifies the efforts to tackle waterlogging, heat stress, drought and the impact of urban flooding.

As a nationally developed process, the Delta Plan is structured around a set of seven ambitions, and aims to facilitate local partnerships between public authorities, businesses and communities. These partnerships are starting to map out the vulnerabilities, draw up a strategy and implementation agenda based on local risk dialogues, while maximising opportunities for linkage. The implementation will be facilitated by a combination of encouragement and regulation.

2.4   MERGING OF WATER MANAGEMENT AND URBAN DESIGN

Solving the urban water assignment is centred on "living with and making space for water" and is both an engineering as well as a design challenge. We have to consider optional alternatives during the design process by linking the water assignment (e.g. required water storage) to the ecological services metrics (e.g. system and site scale measures). The illustration in Fig. 2.5 distinguishes the following four quadrants:

1. Urban analysis in layers—city (use) typologies
2. Water systems approach—cities as water catchments

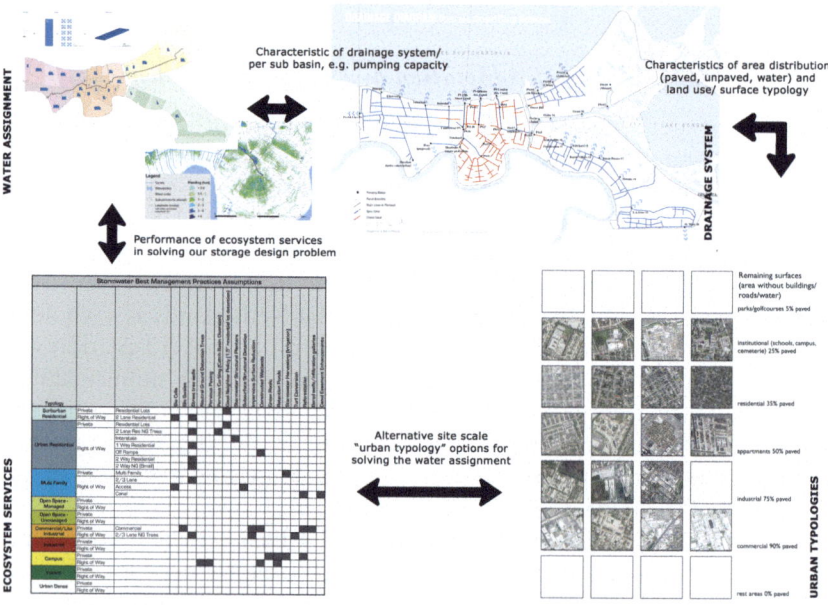

Fig. 2.5   Linking the water assignment to the ecological services metrics

3. The water assignment
4. Green infrastructure—cities providing ecosystem services

2.4.1 Urban Analysis in Layers—City (Use) Typologies

In 1998, a stratified model that distinguished spatial planning tasks on the basis of the differing spatial dynamics of substratum, networks and occupation patterns—that is, three layers—was introduced in the national debate on spatial planning in the Netherlands. Although using layered models was not a new thing, this model hit a nerve in spatial planning practice, initially on a national level, but later also on provincial and municipal levels. Since 1998, this "layers model" has developed into an approach to spatial planning and design: the Dutch layers approach (Fig. 2.6).

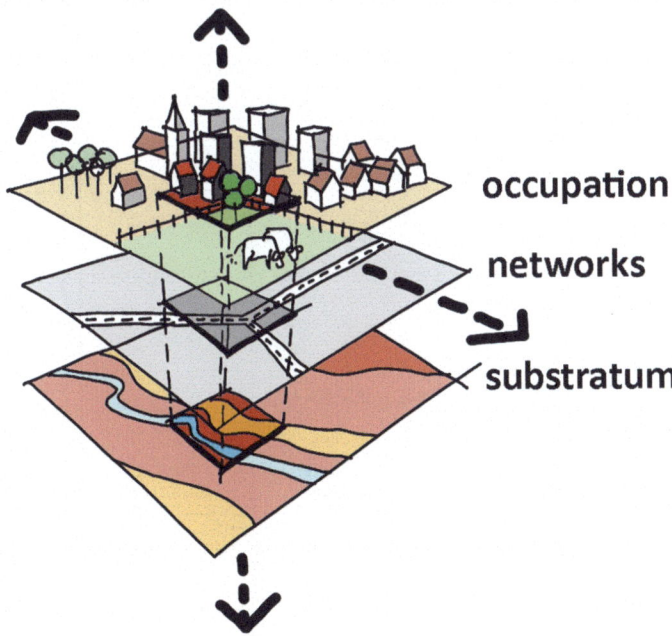

Fig. 2.6 Dutch layers approach to spatial planning and design. (Reproduced with permission from Dauvellier/MIRUP en www.ruimtexmilieu.nl)

2.4.2   Water Systems Approach—Cities as Water Catchments

One of the layers to spatial planning and design is the water system, both on the surface as well as subsurface. Cities are part of and have developed in water catchments. Understanding both the urban and natural water system is vital to propose and evaluate any spatial intervention or new development (Fig. 2.7). After the middle of the nineteenth century, towns and cities expanded to accommodate growth in industrial activity and population. Many waterways lost their infrastructure function and were filled in. During the twentieth century, the urbanisation pattern became linked to the new motorway network and water stopped contributing an organising role to urban development, whilst hydraulic engineering expertise made it possible to protect the land from flooding.

2.4.3   The Water Assignment

The water system analysis can be captured in a water balance or hydrological model. The starting point for the design of any urban water system is an overview of all the functions and functionalities the water system—surface water and groundwater—has to fulfil, now and in a sustainable future.

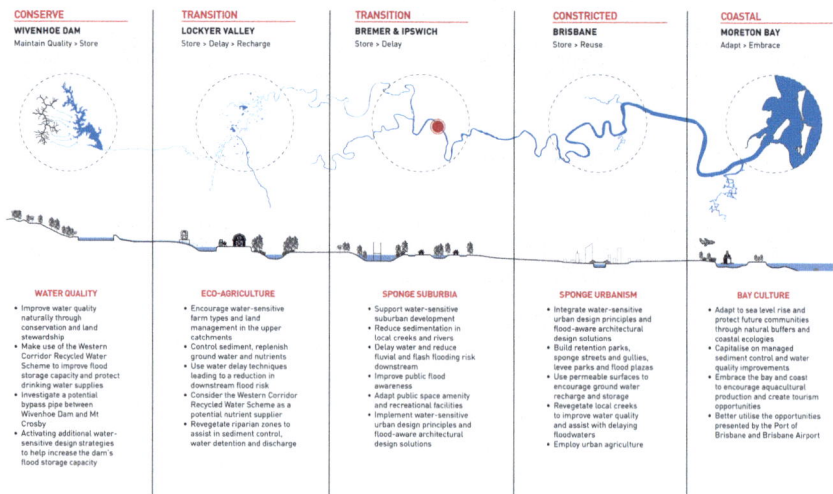

Fig. 2.7 Fluvial transect—cities as water catchments. (Reproduced with permission from James Davidson Architect 2017)

These functions and functionalities are the basis for the design standards that are to be applied. Such an overview of functions and functionalities—ecosystem services, so to say (Sect. 2.4.4)—is to be composed for all the five sorts of urban water; that is for surface water, groundwater, wastewater, storm water and runoff water, as well as for the urban soil and subsurface.

Storage and discharge are exchangeable. Storm water runoff that we cannot discharge needs to be stored temporarily in the system, and we will have to discharge any runoff we do not have capacity to store; we will have to handle this water either way. The required storage capacity (or water assignment, Fig. 2.8) does not only depend on runoff intensity, but also on discharge capacity. As designers, we want to understand the relationship between the required storage capacity and discharge capacity.

Besides increasing discharge and creating more storage capacity, the runoff intensity and volume can be reduced. Runoff intensity depends on the design and construction of the buildings, streets, gardens and so on. Normally storm water will drain quickly from roofs and streets to canals and ponds via the storm sewers; the delay will be no more than

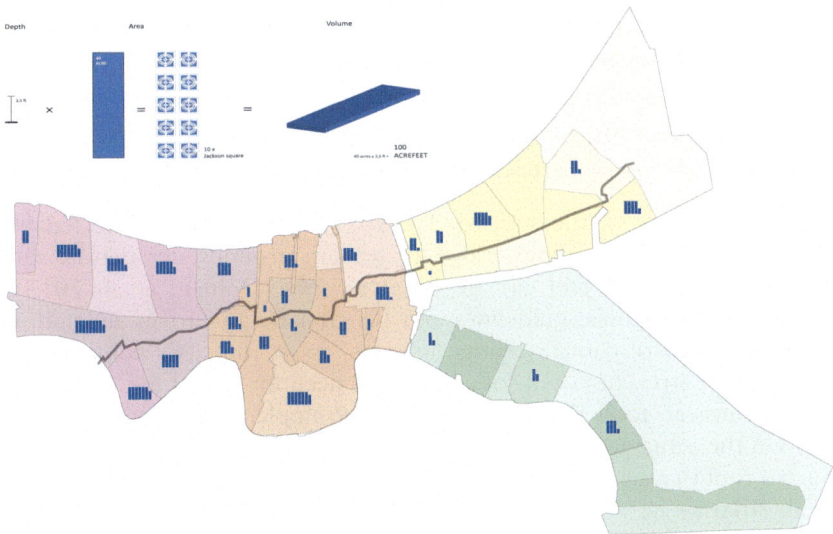

Fig. 2.8 Water assignment per subbasin or neighbourhood (Waggonner et al. 2014)

5–15 minutes, and runoff losses could be 10% or less. But if we could divert the storm water to an infiltration facility via the urban groundwater—in many cases a subsurface drainage system—the delay would be hundreds or even a thousand times larger.

2.4.4   Green Infrastructure—Cities Providing Ecosystem Services

The design solutions to solve the water assignment are elaborated in the city planning and in the suggested urban vision (statement) projects as well as demonstration (pilot) projects. Some practical measures and optional alternatives, both on system and site scale, that can be mentioned, are: (re) use of water, surface drainage (delay), retention (streets, parks, etc.), infiltration (swales, etc.), green infrastructure, storage surface water, adaptive and flood-proof building.

Plenty of WSUD elements exist to reduce runoff and create numerous alternative options for solving the storage design problem (Dolman et al. 2013). These optional alternatives (Table 2.1) have to be addressed during the design process by at least considering separately a fast surface and piped runoff component and a slow runoff component through the soil/subsurface drainage system. It should be considered to incorporate these best practice guidelines or WSUD elements in a manual.

2.5   TESTING OF WSUD INTERVENTIONS IN CLIMATE RESILIENT PATHWAYS

The performance of the suggested WSUD interventions is affected by land use or urban typology, slope of terrain, groundwater depths and dynamics, soil type, and climate characteristics (Dolman et al. 2013). Currently, most tools, guidelines and benchmarks for urban adaptation raise awareness of climate change impacts, assess the city's vulnerability (e.g. climate stress test) and/or address the need for adaptation at a policy level. However, tools that have the ability to implement adaptation solutions in the actual urban planning and design practice seem to be missing. Such a tool that fills this gap is the Adaptation Support Tool (Van de Ven et al. 2016).

This Adaptation Support Tool (AST) has been developed to support local policymakers, planners, designers and practitioners in defining the

Table 2.1 WSUD elements and key selection characteristics

User	Fact Sheet number	Fact Sheet	Stormwater Quality	Stormwater Retention	Wastewater	Aesthetic value	Small	Medium	Large	Broad
Household	1.	Water sensitive homes	–	–	–	–	•			
	2.	Household rainwater tanks	✓	✓	–	–	•			
	3.	Sizing a rainwater tank	✓	✓	–	–	•			
	4.	Porous paving	✓	✓	✗	✗	•	•	•	•
	5.	Site layout and landscaping	✓	✓	✗	✗	•			
Developers, Council planners, architects, engineers	6.	Water conservation initiatives	–	–	–	–	•	•	•	
	7.	Waterway rehabilitation	✓	✗	✗	✓		•	•	•
	8.	Rainwater tanks	✓	✗	✗	✗		•	•	•
	9.	Gross pollutant trap	✓	✓	✗	✓?		•	•	•
	10.	Sedimentation (settling)	✓	✓	✓?	✓	?	•	•	•
	11.	Ponds and lakes	✓	✓	–			•	•	•
	12.	Vegetated swales and buffer strips	✓	✗	✗	✓	•	•	•	•
	13.	Raingardens	✓	✗	✗	✓	•	•	•	
	14.	Raingarden tree pit	✓	✗	✗	✓	•	•	•	
	15.	Surface wetlands	✓	✓	✓?	✓	?	•	•	•
	16.	Subsurface flow wetlands	✓?	✓?	✓?	✓	•	•	•	
	17.	Suspended growth biological processes	✗	✗	✓	✗	?	•	•	•
	18.	Fixed growth biological processes	✗	✗	✓	✗	?	•	•	•
	19.	Recirculating media filter	✗	✗	✓	✗	?	•	•	
	20.	Sand and depth filtration	✗	✗	✓	✗	?	•	•	
	21.	Membrane filtration	✗	✗	✓	✗	?	•	•	•
	22.	Disinfection	✗	–	✓	✗	?	•	•	•

✓ = Primary purpose ✓? = Some impact but not primary purpose ✗ = Does not contribute
– = not applicable ? = possibly applicable • = applicable

Reproduced with permission from City of Melbourne (2012)

programme of demands and setting adaptation targets. The AST can be used in design workshops, to support dialogue among stakeholders on where and how each ecosystem-based adaptation measure can be applied. The tool provides quantitative, evidence-based performance information on (cost)effectiveness of adaptation measures regarding climate resilience and co-benefits (Fig. 2.9). The AST includes a toolbox of 62 blue, green and grey measures for ecosystem-based adaptation —also called blue-green infrastructure (such as green roofs, bioswales, porous pavement and water squares)—a selection assistant for ranking based on their applicability and an assessment tool to estimate the effectiveness of applied adaptation measures. The performance of the measures has been estimated in the

Fig. 2.9 Screen components of the Adaptation Support Tool. Left on the touch screen is the ranked list of 62 adaptation measures. Selected measures are planned in the project area (middle). At the right side the AST dashboard, showing the resilient performance of the total package of measures and of each active measure. Shown is the application of the AST in Beira, Mozambique (Van de Ven et al. 2016)

AST for a number of key metrics, for example, created storage capacity, frequency reduction of the normative runoff, heat stress reduction, extra groundwater recharge, water quality effects, costs and additional benefits. Performance is determined for each adaptation measure separately as well as for the total package of measures.

Collaborative spatial planning isn't limited to retrofitting attractive solutions and the effectiveness of single adaptation measures. It's also about the collective; green-blue urban grids make cities sustainable, resilient and climate-proof. Working towards a liveable and attractive green-blue city has to be considered at the forefront of a climate resilient pathway. By strengthening green infrastructure and giving water more space in both the public and private domains, cities have the potential to grow into the blue-green cities. Cities can therefore focus on a blue-green design based on the three design scales included in Table 2.2.

Table 2.2 Design scales for a blue-green city (Dolman 2019)

Blue-Green design principle	Spatial scale	Who?
Sufficient urban "sponges" for detaining (using), retaining or delaying rainwater	Buildings, streets and neighbourhoods	City together with its residents and actors
Blue-green city network on which "sponges" can drain excess water and in which discharge and storage take place	Neighbourhoods, districts and city	City together with water authority
Emergency valves for the blue-green network and overflow areas where water can temporarily go in extreme situations	City, region and delta	City together with water authority, regional authority and neighbouring cities

2.6 Application in Practice: Case of Hoboken, USA

The response to the flood disaster in New York and (urban) New Jersey caused by Hurricane Sandy in 2012 resulted in a set of approaches proposing a transition towards water resilient cities on the USA's east coast, including the urbanised areas around the Hudson River, such as the New York metropolitan region. In the wake of the flooding caused by Hurricane Sandy, New York City produced a noteworthy study with a series of concrete measures: "A stronger, more resilient New York". This body of work includes an assessment of the challenges the city faces, how these are expected to increase as a result of climate change and what type of measures can be implemented to ensure a more resilient New York City. The report has a strong scientific basis and has a focus on local, small-scale, green infrastructure measures. The elements presented can contribute to a transition towards a Water Resilient City, although there is still a need for a wider framework and overarching vision. It is too early to see the effect of the implementation of the proposed measures, yet the strategy is a valuable first step.

A design approach to achieve resiliency was initiated on a larger scale in the Sandy-affected region (including New York, New Jersey and Connecticut) by the Hurricane Sandy Rebuilding Task Force. As part of their activities, the Task Force invited the world's leading engineers and architects to participate in a design competition: "Rebuild by Design". Instead of just building back to a pre-Sandy condition, the emphasis was put on building back smarter. Multidisciplinary teams engaged with local stakeholders to analyse the region and propose new concepts to deal with new climate extremes. One of the tracks included approaches for high-density urban areas.

As part of the Rebuild by Design competition, the team led by OMA proposed a Comprehensive Water Strategy for "high-density urban environments", with the City of Hoboken (New Jersey) as a demonstration project area (see Fig. 2.10). Hoboken lies on the west bank of the Hudson River in the New York metropolitan area, directly across from Manhattan. It is the fourth-densest city in the USA, and includes a major transportation hub for the region. Due to its high-density urban environment and low elevation, it is also very vulnerable to both flash floods and storm surges. Its single watershed, single jurisdiction, and combination of high-impact factors (high density, value, influence and potential) lends itself to creating a multifaceted solution that both defends the city as a whole and enables the development of commercial, civic and recreational amenities.

The proposed comprehensive urban water strategy deploys programmed hard infrastructure and soft landscape for coastal defence (resist); policy recommendations, guidelines and urban infrastructure to slow rainwater runoff (delay); a circuit of interconnected green infrastructure to

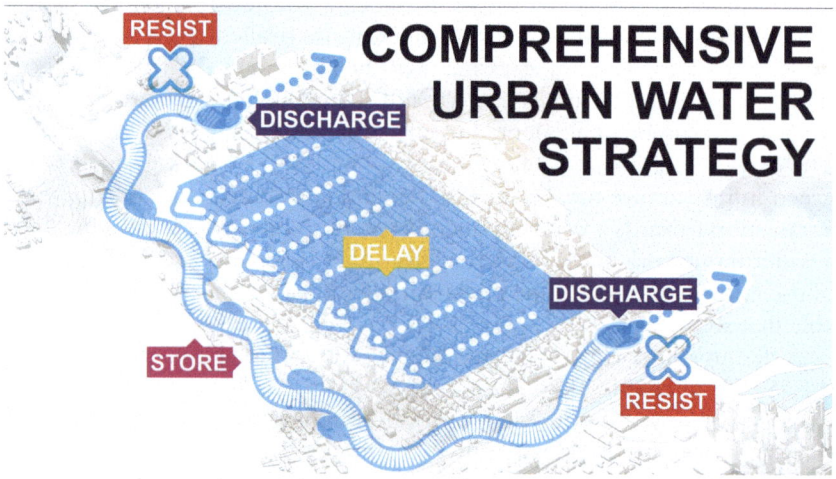

Fig. 2.10 Hoboken comprehensive urban water strategy. (Reproduced with permission from OMA and Royal HaskoningDHV 2014)

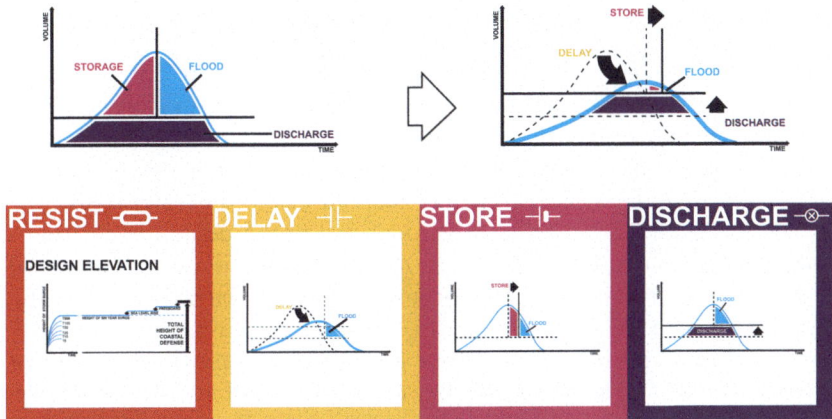

Fig. 2.11 "Resist – Delay – Store – Discharge" approach. (Reproduced with permission from OMA and Royal HaskoningDHV 2014)

store and direct excess rainwater (store); and water pumps and alternative routes to support drainage (discharge). The proposed set of measures that make up the comprehensive urban water strategy can be summarised by their actions: "Resist-Delay-Store-Discharge" (see Fig. 2.11). The objectives of this manifold strategy are to manage water, for both disaster and for long-term growth, and the delivery of multiple benefits—including civic, cultural, recreational and commercial amenities—that enhance the quality of the built environment.

The strategy addresses many aspects of a water resilient city: a comprehensive approach to flood risk; a coalition of stakeholders and collaborative funding framework; an umbrella of communication and education; and integrated multifaceted design solutions. In 2015, the "Hoboken comprehensive urban water strategy" was awarded with a role model city certificate of the UNDRR "Making Cities Resilient Campaign". The initiatives in urban New York and New Jersey show great ambition and the potential to transform vulnerable megacities into leading examples of water-sensitive communities. To what extent the implementation will follow through with the outlined approach remains uncertain.

2.7   DISCUSSION AND CONCLUSIONS

The awareness in cities is growing that water challenges are driving a need for a change in the way we develop our urban areas. Both water and climate change may trigger the next financial crisis (World Economic Forum 2014), in particular when it is considered how little tangible action has been taken to address climate change adaptation and the water crisis.

The integration of water management and urban design, like sustainability and resiliency, is very complex. Although water is a crucial aspect of resilience, the city strategies are not just focused on water. It is about the capacity of individuals, communities, institutions, businesses, even systems, within a city to survive, adapt and grow no matter what kind of chronic stresses and acute shocks they experience. And water is a connecting challenge in making cities resilient. The water issue then becomes a way of looking at cities and at how to make them climate- and future-proof.

To improve climate resilience and reduce vulnerability to water extremes, research shows that there are five pillars that need to be strengthened (Chap. 1). In bridging water management and urban planning, experience has been gained in these capacities: threshold capacity, coping capacity, recovery capacity, adaptive capacity and transformative capacity. These five capacities should be considered as incremental or constructive, as illustrated in the Resist, Delay, Store, Discharge (RDSD) strategy as part of the Rebuild by Design competition in 2013 (Sect. 2.6). The "Resist" part of the strategy covers the threshold capacity. And where the "Delay, Store, Discharge" part builds adaptive capacity to climate change and to enhance liveability, it also relates to improving preparedness or coping capacity and recovery capacity.

An adaptive city is a rather reactive terminology that looks at how cities can build (engineered) resistance against future shocks and stresses, such as from climate change and urbanisation. It is about lasting and making it through a crisis rather than trying to stop the development that causes the crisis. Urban resource consumption, waste disposal and loss of forest and natural land are widely seen as the root cause of many of the world's environmental problems. Because so much damage has already been done to the world's ecosystems, and solutions need to be found to reverse it, we need to start thinking of regenerative urban ecosystems. This requires investing in the transformative capacity. Also, adaptation to climate change is a transition, for example, illustrated by the Urban Water Management Transition Framework (Figs. 2.1 and 2.2). The latter aims for a "*Water*

Sensitive City which combines physical infrastructure, such as water-sensitive urban design (WSUD), with social systems (e.g. governance and engagement) to create a city where the infrastructure and systems enhance the connections people have with water and improve quality of life".

Following on from frameworks like the Water-Sensitive City (Brown et al. 2008), the International Water Association (IWA) developed the principles for Water-Wise Cities (IWA 2016). The 17 principles are grouped into four levels of action: (1) Regenerative Water Services, (2) Water-Sensitive Urban Design, (3) Basin Connected Cities, and (4) Water-Wise Communities (Fig. 2.12). These principles can help city leaders ensure that everyone in their cities has access to safe water and sanitation. One of the aims is to ensure that water is integrated in planning and design in cities to provide increased resilience to climate change, liveability, efficiencies and a sense of place for urban communities. The ultimate goal of these principles is to encourage collaborative action underpinned by a shared vision, so that local governments, urban professionals and individuals actively engage in addressing and finding solutions for managing all waters of the city.

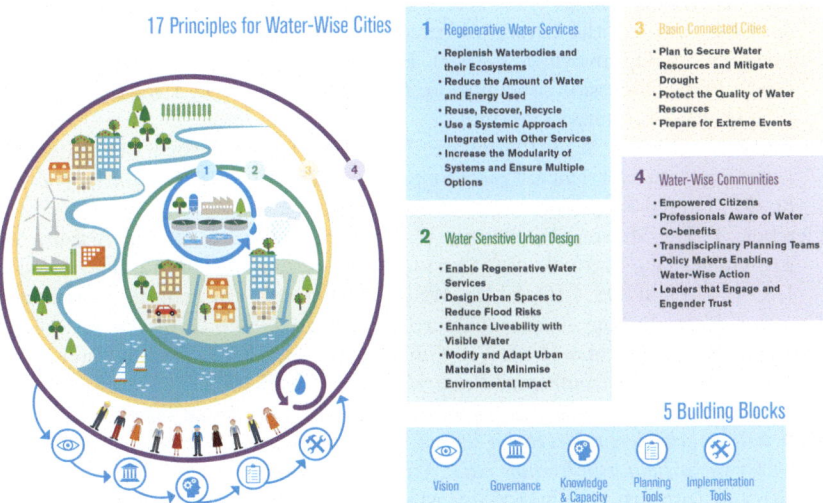

Fig. 2.12 Principles for Water-Wise Cities. (Reproduced with permission from IWA 2016)

This complexity brings opportunities. Hopefully the momentum around the Paris climate change agreement will provide an opportunity to go beyond action on greenhouse gas emissions to achieve progress on wider issues such as planning for water resilience and adaptation to climate change. Cooperation is going to be needed for progress on urban water resilience to be achieved. If people are able to take this advice on board, they can contribute to the change that is needed for a resilient future.

REFERENCES

Brown, R., Keath, N., & Wong, T. (2008). *Transitioning to water sensitive cities: Ensuring resilience through a new hydro social contract.* 11th international conference on urban drainage, Scotland, UK.

Brown, R., Rogers, B., & Werbeloff, L. (2016). *Moving toward Water Sensitive Cities: A guidance manual for strategists and policy makers.* Melbourne: Cooperative Research Centre for Water Sensitive Cities.

City of Melbourne. (2012). *Applying the model WSUD guidelines.* Melbourne.

Davidson, J. (2017). *The water futures book, James Davidson architect.* Brisbane.

Dolman, N. (2019). Adaptation strategy for Zwolle – Towards a liveable and attractive blue-green city. *Water Solutions, 4,* 18–21.

Dolman, N., & Ogunyoye, F. (2018). How water challenges can shape tomorrow's cities. *Proceedings of the ICE – Civil Engineering, 171*(6), 22–30, special issue on "Cities of the Future", United Kingdom.

Dolman, N., & Ogunyoye, F. (2019). How water challenges can shape tomorrow's cities – Discussion contribution. *Proceedings of the ICE – Civil Engineering, 172*(1), 13–14, United Kingdom.

Dolman, N., Savage, A., & Ogunyoye, F. (2013). Water-sensitive urban design: Learning from experience. *Proceedings of the ICE – Municipal Engineer, 166*(2), 86–97, United Kingdom.

Fletcher, T. D., Shuster, W., Hunt, W. F., Ashley, R., Butler, D., Arthur, S., Trowsdale, S., Barraud, S., Semadeni-Davies, A., & Bertrand-Krajewski, J.-L. (2014). SUDS, LID, BMPs, WSUD and more – The evolution and application of terminology surrounding urban drainage. *Urban Water Journal,* 1–18.

Hoekstra, A. Y., Buurman, J., & Van Ginkel, K. C. H. (2018). Urban water security: A review. *Environmental Research Letters, 13,* 053002.

International Water Association. (2016). *Principles for Water Wise Cities – For urban stakeholders to develop a shared vision and act towards sustainable urban water in resilient and liveable cities,* United Kingdom.

Ministry of Infrastructure and Water Management (Netherlands). (2017). *Delta plan on spatial adaptation.* The Hague.

Ministry of Transport, Public Works and Water Management (Netherlands). (2000). *A different approach to water – Water management policy in the 21st century.* The Hague: Reinders Partners.

Ministry of Transport, Public Works and Water Management (Netherlands). (2009). *National Water Plan – Multi-layer safety as a central concept.* The Hague.

OMA and Royal HaskoningDHV. (2014). *Rebuild by design [Resist, Delay, Store, Discharge] project – A comprehensive urban water strategy for the Hudson Waterfront.* Hoboken, State of New Jersey, United States of America.

Šakić Trogrlić, R., Rijke, J., Dolman, N., & Zevenbergen C. (2018). *Rebuild by Design in Hoboken: A design competition as a means for achieving flood resilience of urban areas through the implementation of green infrastructure.* Accepted manuscript in MDPI (open access) Journal of Water, Switzerland.

Tjallingii, S. (1996). *Ecological conditions. Strategies and structures in environmental planning.* IBN Scientific contributions 2. Wageningen, IBN-DLO. 320 PP. ill. ISBN 90-801112-3-6. Doctoral Delft University of Technology thesis.

USEPA. (2003). Protecting water quality from urban runoff 847-F-03-003. http://www.epa.gov/npdes/pubs/nps_urban-facts_final.pdf

Van de Ven, F. H. M., Snep, R. P. H., Koole, S., Brolsma, R., van der Brugge, R., Spijker, J., & Vergroesen, T. (2016). Adaptation Planning Support Toolbox: Measurable performance information based tools for co-creation of resilient, ecosystem-based urban plans with urban designers, decision-makers and stakeholders. *Environmental Science & Policy, 66,* 427–436.

Waggonner, D., Dolman, N., Hoeferlin, D., Meyer, H., Schengenga, P., Thomaes, S., Van den Bout, J., Van der Salm, J., & Van der Zwet, C. (2014). New Orleans after Katrina: Building America's water city. *Proceedings of Built Environment, 40*(2), 281–299.

WEF (World Economic Forum). (2014). Global Risks Report 2014. WEF, Cologny, Switzerland.

Zevenbergen, C., Cashman, A., Evelpidou, N., Pasche, E., Garvin, S., & Ashley, R. (2012). *Urban flood management.* Boca Raton: CRC Press.

CHAPTER 3

Climate Resilient Urban Retrofit at Street Level

Jeroen Kluck and Floris Boogaard

Abstract For a successful transition of existing urban areas towards climate resilience, it is essential that macro level policies and conceptual approaches find their way to every street and neighbourhood through urban retrofit projects. This chapter illustrates how technology, research and urban design can assist and inform this transformation process towards climate resilient streets. It describes three important steps for improving the uptake of climate resilient design in practice:

1. Implementation of design guidelines for climate resilience
2. Show that climate resilient designs are possible, not complicated and affordable
3. Show international examples from practice

J. Kluck
Urban Water, Amsterdam University of Applied Sciences,
Amsterdam, The Netherlands

F. Boogaard (✉)
Urban Water, Hanze University of Applied Sciences Groningen,
Groningen, The Netherlands
e-mail: f.c.boogaard@pl.hanze.nl

© The Author(s) 2021
R. de Graaf-van Dinther (ed.), *Climate Resilient Urban Areas*,
Palgrave Studies in Climate Resilient Societies,
https://doi.org/10.1007/978-3-030-57537-3_3

45

These steps are presented by (a) a book of example climate resilient designs for characteristic urban typologies with life cycle costs analyses and (b) a worldwide climate mapping platform presenting many implemented projects. With guidelines *implemented* and the knowledge that climate adaptation is *affordable* and has been done before it should become evident that most urban reconstruction projects can and should be climate resilient projects.

Keywords Climate resilient design • Urban typologies • Climate adaptation platform

3.1   Introduction

In order to stay livable, cities need to adapt to the changing climate. In fact cities have been designed and built for the climate of the past and are not prepared for the coming weather conditions (GCA 2019). The frequency and damage of storm water flooding will increase as the urban water system is not designed for the intensification of extreme rainfall events (IPCC 2019). More periods of drought will increase damage on green and infrastructure. And increase of heatwaves will result in a range of problems from an unattractive urban environment to health problems (Klok and Kluck 2018). In addition to climate change impacts, cities are changing—mostly growing, generally amplifying the above problems because green and pervious areas are replaced by impervious areas, reducing water storage, discharge, infiltration and evaporation, and reducing cool green spots (Majidi et al. 2019). Because of the expected problems, Dutch municipalities have the task to take climate adaptation into account and to use all urban development projects and retrofitting or refurbishing of existing urban areas to step by step make all cities climate proof (Deltaprogramma 2014). But for a successful transition of existing urban areas towards climate resilience, it is essential that macro level policies and conceptual approaches find their way to every street and neighbourhood in the urban retrofit projects.

The Dutch Association of Municipalities has agreed with the national government that from 2020 on all municipalities will reconstruct urban areas in a climate resilient way. As most urban streets are reconstructed only once in 30 years' time, all existing urban areas can be climate resilient by 2050. A timeline has been set up for all municipalities to do climate risk assessments, risk dialogues and finally start implementing measures. To facilitate this process, knowledge on climate resilient design is being shared on a national climate services portal (CAS 2020) and many consultancies have picked up this opportunity to show their expertise and offer climate resilience studies to municipalities.

In practice municipalities have so many tasks to fulfil when retrofitting an area that the extra task of climate adaptation is easily dropped. Professionals at frontrunning Dutch municipalities expressed the feeling that many solutions are available but that the difficulty is how to convince others to implement them.

Both Amsterdam University of Applied Sciences and Hanze University of Applied Sciences Groningen have interdisciplinary knowledge centres focusing on urban climate adaptation with and for municipalities. In recent years they have in close contact with Dutch municipalities picked up the issue of how to retrofit urban areas in a climate resilient way. Based on the cooperation with those municipalities the authors of this chapter have defined three important steps for improving the uptake of climate resilient design in practice.

1. Implementation of design guidelines for climate resilience
2. Show that climate resilient designs are possible, not complicated and affordable
3. Show international examples from practice

With guidelines *implemented* and the knowledge that climate adaptation is *affordable* and has been done before it is increasingly accepted that urban reconstruction projects can and should be climate resilient projects.

3.2 GUIDELINES

In the Netherlands there are clear and very high standards for coastal and river flood safety (e.g. once in 5,000 to 12,000 years). For urban pluvial flooding and heat stress, such *binding* standards have not been set. Currently urban designers/developers/managers who are dealing with climate resilient design in the Netherlands are busy discussing on design goals, thresholds and design rules. There is no national *binding* standard for the acceptable levels of pluvial flooding or heat in the urban space.

Municipalities have to make their own choice, and they are experimenting with different levels of safety. Several municipalities cluster together to assist each other and exchange knowledge. Some municipalities or groups of municipalities have recently adopted some design rules or guidelines aiming at climate resilience in new developments and reconstruction projects or they are experimenting with such design rules.

Guidelines would certainly be an important enabling factor in making every project for reconstruction or retrofitting an urban area a climate resilient project. With clear design guidelines, municipalities can take into account climate resilience in their designs. However, the problem in the Netherlands is that for many of the climate risks there is no consensus at all on what would be sufficiently climate resilient. Authorities first want to know how large the risks are, what are the costs for possible solutions and who is going to pay for it. As storm water flood risk and damage, and costs of interventions can vary greatly per municipality, local municipalities do not want to be instructed what interventions to choose. At present, some agreement seems to arise slowly on storm water flooding standards. For example, the municipality of Amsterdam requires that rainfall of 60 mm/hour should not lead to flooding of buildings (Kluck et al. 2015). On urban heat, progress is made as well, but this is even more difficult as the goals for a climate resilient urban space are very unclear. Unlike storm water flooding (where it is possible to estimate which rainfall intensity will cause flooding in a building), for urban heat there is no clear threshold above which urban space is unlivable. The higher the temperature during a heatwave the more people will suffer.

Without clear guidelines many reconstruction projects will lack the required direction to achieve climate resilience. Fortunately, quite some municipalities are already using or experimenting with their own guidelines: for instance Amsterdam, as mentioned, and Eindhoven with a prescribed percentage of green space for urban projects.

3.3   Standard Solutions

In request of professionals at frontrunning Dutch municipalities, who expressed the feeling that many solutions for climate resilient are available but that the difficulty is how to convince their colleagues, a project has been set up to create and compare alternative climate resilient designs to the standard (non-climate resilient) design.

The approach of this project was to use characteristic urban typologies as a starting point. Ten case studies within these typologies were selected. And for each case study, a standard design was compared to several more flood resilient variant designs. To compare the variants, for each variant, costs and benefits were estimated.

3.3.1 Urban Typologies

Street design in the Netherlands is often based on a particular philosophy of its time. Ideas and technologies that were available at the time of constructing are captured in the authentic details of these streets, such as the size of the houses, gardens, public space for greens and playgrounds, the width of the streets and the architecture of the buildings (Kleerekoper 2016). Kleerekoper (2016) also describes a set of neighbourhood typologies for urban climate adaptation (Fig. 3.1 and Table 3.1). The typological variants give direction to the approach to combat more extreme climate effects. For instance, the abundance of public space in post-war neighbourhoods can easily be employed for climate adaptation, whereas in the dense urban housing blocks and pre-war blocks, underground solutions

Fig. 3.1   (a–c) Three of the Dutch neighbourhood typologies (Kluck et al. 2018)

Table 3.1 Dutch neighbourhood typologies, based on (Kleerekoper 2016)

Dutch neighbourhood typology	Period	Features
Urban city block	Before 1930	No front garden nor green skirting, 4–5 layers
Pre-war city block	1900–1940	Occasional front garden, 3–4 layers, wider streets than urban blocks and occasional green skirting
Garden village	1910–1930	Spacious front and back garden, 2–3 layers, ample parking space, 1930s' architecture, limited public green and rarely street trees
Working-class neighbourhood	1930–1940	No front garden, little public green, 2–3 layers, single-family units
Low-rise post-war garden city	1945–1955	Open building block with ample green, 2–3 layers, single-family units
High-rise post-war garden city	1950–1960	Open building blocks with ample green, 4–6 layers, apartments, storage on the ground level
Post-war neighbourhood	1940–1990	Front and back garden, 2–3 layers, single-family terraced houses, semi-detached or detached
Community neighbourhood	1975–1980	Single-family unit with front- and back garden, meandering street pattern, courtyards, wide green skirting around the neighbourhood
High-rise city centre	1960–present	More than 10 layers in grid formation
Suburbanization— Vinex	1990–2005	Single-family unit, terraced, semi-detached or detached apartments

are more important. The structure of garden cities offers space for swales to absorb heavy rainfall locally. Knowledge of the neighbourhood typology, gradient (flat or sloping), type of soil and the groundwater level enables us to give a reliable projection of the possibilities and effectivity of local climate adaptation. The typologies apply to many of the streets and neighbourhoods across the Netherlands. In every country common typologies can be determined to present climate adaptations that generally fit in. Throughout Europe typologies will vary strongly, especially from North to South due to difference in climate. Northern countries tend to have more spacious streets to allow sunlight entering the houses during winter.

3.3.2   Ten Examples with Variants

Ten examples of streets in neighbourhoods that are typical and representative of the Dutch infrastructure (e.g. the urban city block, post-war garden cities and community neighbourhoods) were selected. They include flat and sloping surfaces, and differences in soil permeability and groundwater levels, and represent Dutch situations. The ground water tables are generally high, the Dutch soil is predominantly sandy with clay or peat (Kluck et al. 2018), and the Netherlands has a predominantly flat surface with some sloping areas in the eastern- and southern regions. Six of the ten examples were in a flat area. The other four were in a mild hilly area.

For each example, a traditional (standard) design and three climate resilient variants have been worked out. Variant 0 is the traditional design, whereas variants 1–3 are more climate resilient to extreme rainfall (Fig. 3.2). Additionally, some examples include a particularly green variant.

Variants were tested on their sensitivity to flooding, assuming that flood damage occurs when water enters the houses. For that purpose we calculated for different volumes of extreme rainfall in one hour if water would enter the houses. These different extreme volumes of rainfall are linked to estimations of frequency of occurrence.

3.3.3   Costs and Benefits

Next the lifetime costs and benefits of each variant were estimated by a methodology described in Kluck et al. (2018). This methodology includes flood damage costs and life cycle costs (i.e. construction and maintenance costs) of, for example, sewer systems and permeable paving. Furthermore, it includes variation in the lifespan of particular refurbishments. Calculations were based on the cost ratios of the Dutch Sewer Guidelines (Stichting Rioned 2015) as well as empirical evidence provided by individual municipalities. Kluck (2017) gives more background documentation on the methodology and a calculation sheet. All investments have been based on a 100-year statistical return period to provide a realistic comparison of variants with differences in maintenance costs. This should provide a deeper insight into the financial consequences of various options to allow policy makers, designers, administrators and other experts to make well-informed decisions.

Climate resilient refurbishment of public space has certain benefits. The calculations include the quantifiable cost-effective measures. The most

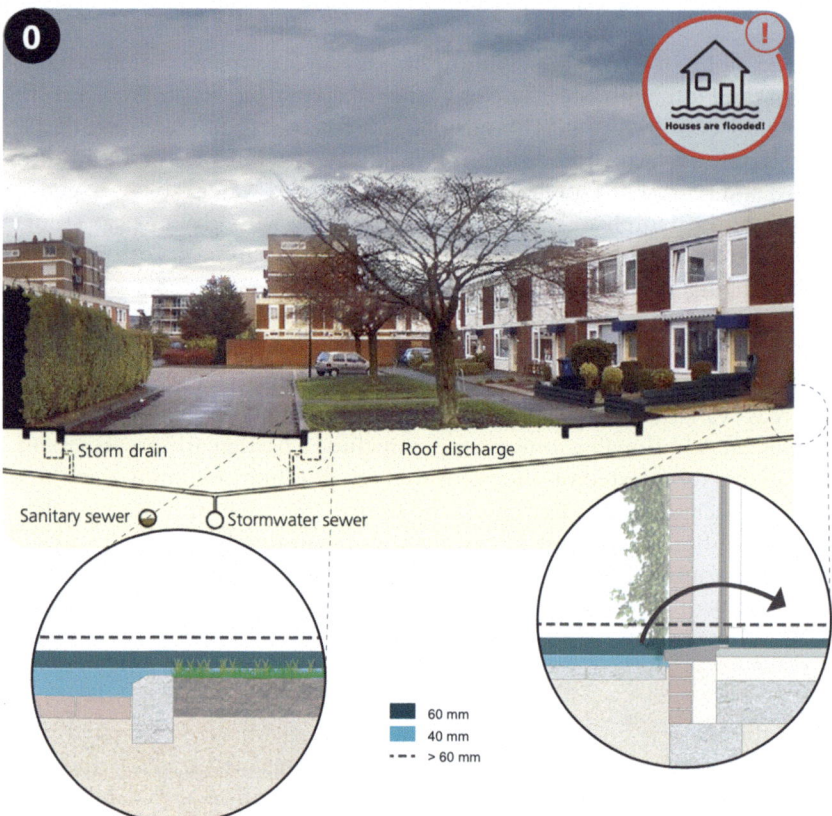

Fig. 3.2 Example of standard design and 3 more climate resilient variants for case 'low rise post war garden city' in flat urban area (Kluck et al. 2018)

important are the reduced flood damage costs which were also expressed in costs per annum based on estimated frequency and the magnitude of the disruption. Other benefits which were calculated are the reduction in maintenance costs of the urban drainage system and—in case of a combined sewer system—reduction in wastewater treatment costs. There are more benefits that were not quantified, such as reduced or delayed drainage to surface water, groundwater recharge, heat stress reduction, and increased water availability for urban green.

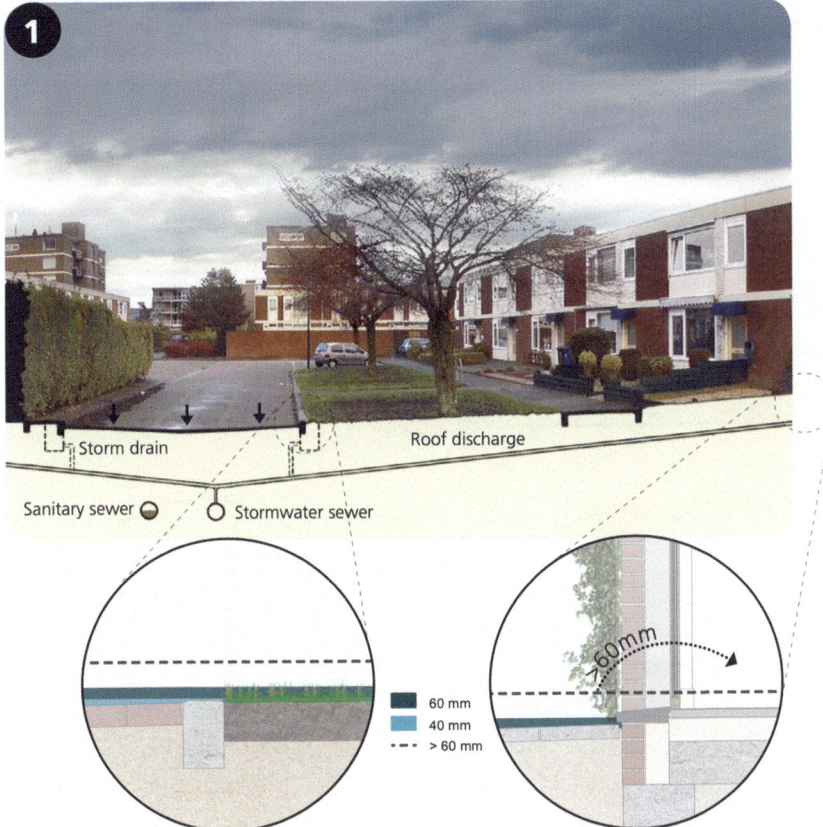

Fig. 3.2 (continued)

In addition to advantages to the water system (such as increased rainwater infiltration), greening public space improves public comfort and health, water quality, and reduction of energy consumption, and it increases biodiversity. These benefits were studied and quantified with the TEEB-city-method ('The Economics of Ecosystems and Biodiversity' Buck Consultants International 2016).

To sum it up: We have compared variants for the costs of maintenance and implementation as well as the benefits of reduced or prevented flood damage and greening. Benefits that we have so far not been able to quantify sufficiently have not been included.

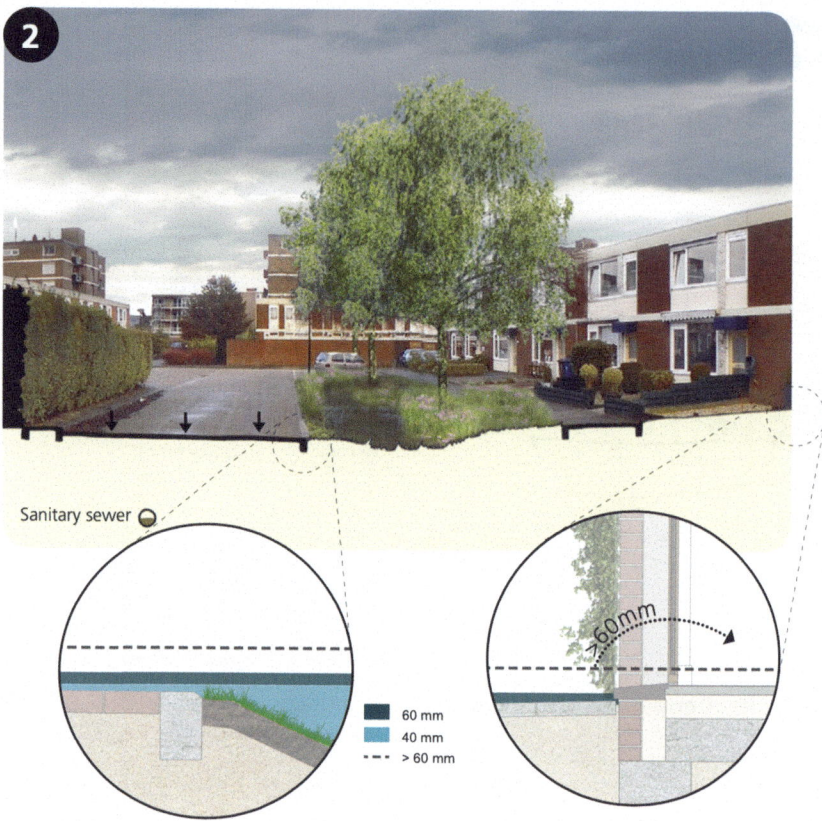

Sanitary sewer

60 mm
40 mm
- - - > 60 mm

Fig. 3.2   (continued)

3.3.4   *Results*

The comparison of the variants shows that the climate resilient variants are not necessarily more costly than the standard situation. Moreover, they are relatively easy to implement when they can piggyback on planned urban refurbishment and maintenance operations. Figure 3.3 presents a comparison of the average yearly lifetime costs for storm water flood damage, construction costs and maintenance costs.

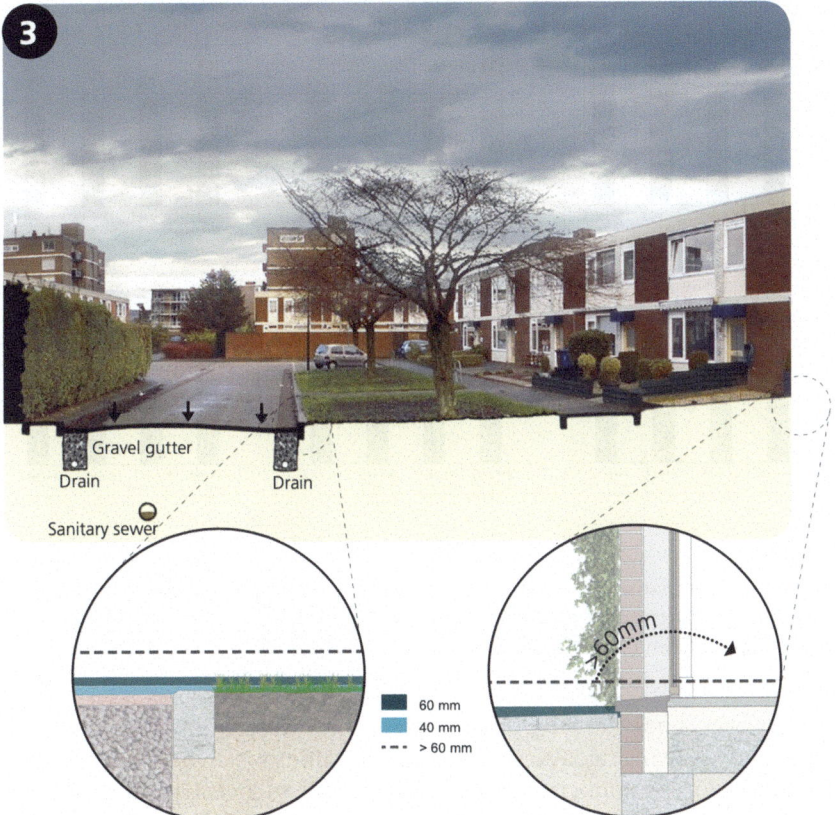

Fig. 3.2   (continued)

The examples with the comparison of costs and benefits have been published in a book of examples (Kluck et al. 2018) targeted at urban professionals. This book shows how ordinary residential streets can be made climate resilient in practice. Explaining that solutions are possible and that the costs are often about the same can help to convince municipalities to choose for a climate resilient design.

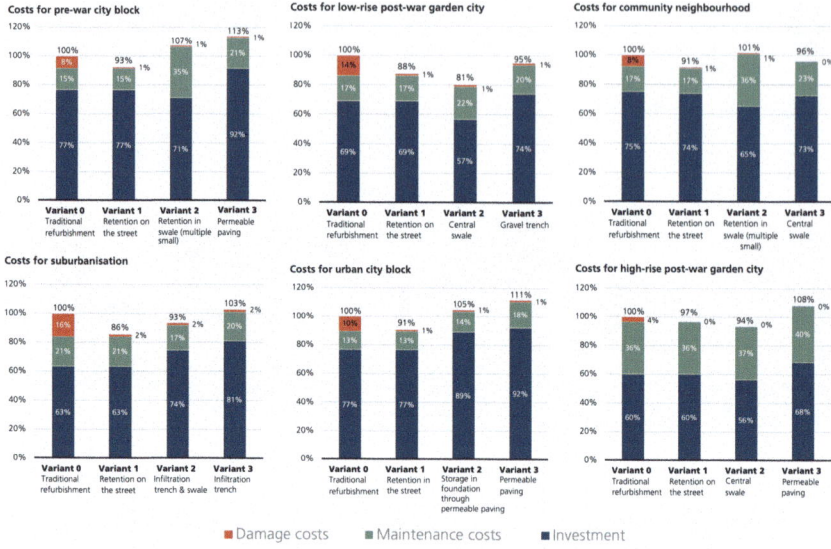

Fig. 3.3 Comparison of lifetime costs (damage, maintenance and construction cost) for rainwater resilient design versus standard design for six urban typology cases adapted from (Kluck et al. 2018)

However, since showing designs of specific interventions might not be sufficiently convincing, also showing interventions which have been implemented before is important. Web-based portals such as ClimateScan (see next paragraph) could also contribute to accelerated uptake of climate resilience measures.

3.4   Existing Solutions Presented in Climate Adaptation Platforms

3.4.1   Introduction to Climate Adaptation Platforms

There is an increasing demand for a collaborative knowledge-sharing on climate adaptation and mitigation. Climate resilient solutions are often already available, but their multiple benefits are often unknown. The aim of most Climate Change Adaptation Platforms is (inter)national knowledge exchange and raising awareness about climate adaptation in urban areas and to promote solutions such as Nature-Based Solutions (NBS),

Fig. 3.4 Global Climate Change Adaptation Platforms as presented during 2nd international climate change adaptation platform workshop in Dublin October 2019 (Climate Ireland 2019)

Best Management Practices, water-sensitive urban design (WSUD) and Sustainable urban Drainage Systems (SuDS).

Climate adaptation platforms are available for several regions or countries around the world offering information to different target groups. Most of the platforms are country or continent specific, and some are global (Fig. 3.4). Examples of Dutch climate adaptation platforms are CAS (2020) and ClimateScan (2020) which offer examples of climate adaptation in the Netherlands. The open-source platform ClimateScan is a global platform and is focussed on street level and will be discussed in this chapter.

3.4.2   ClimateScan Platform

ClimateScan, a web-based international knowledge exchange tool on urban resilience, is a citizen science tool created through 'learning by doing' (Boogaard et al. 2017). Since the implementation in 2014, the platform is in continuous development as more data is uploaded, and improvements are made to respond to feedback from users. In the early stage of ClimateScan, the tool was evaluated by semi-structured interviews in the ClimateScan community with the following recommendations:

Most stakeholders (Dutch municipalities and water authorities) would like to have tools that are (Tipping et al. 2015):

1. interactive;
2. open source;
3. provide more detailed information (location, free photo and film material);
4. link to scientific research outcomes for that specific location;
5. local examples and international examples.

In 2020, ClimateScan has grown into an interactive web-based map application for international knowledge exchange on 'blue-green' projects around the globe. The platform has evolved into a 'solutions-broker'. This role can assist in mobilizing action in the field of climate adaptation. For instance, by creating a network of collaborators and by providing these collaborators with a resource for assistance in their efforts towards achieving resilience. With its existing widespread user base and its diverse portfolio of climate adaptation and mitigation projects, it is a widely used support tool for climate adaptation.

Fig. 3.5 ClimateScan.org platform with around 5000 projects around the world

One of the unique features of ClimateScan is its open-source character which allows users to record and map climate adaptation of their own solutions. The ClimateScan app that can be downloaded in the app store is used for uploading climate adaptation examples all over the world (Fig. 3.5). Unlike other knowledge dissemination platforms, ClimateScan is global in its approach and scope. Most other databases are either restricted in their geographical scope or they have a singular focus on one kind of adaptation solution. The diversity in topics and its global outreach has enabled it to gain a widespread user base, becoming particularly popular among young practitioners and academicians.

Climate scan collects climate adaptation locations from all over the world (Table 3.2). The global platform can be used as a first step to collect data by the means of citizen science and share the knowledge on realized climate adaption measures of that region and compare this to other parts of the world. ClimateScan focuses mainly on the topics surrounding the areas of urban resilience, climate proofing and climate adaptation. The main objective of this interactive international open access platform is knowledge exchange on climate adaptation projects through the platform itself and the connected social media channels as twitter and Facebook.

Due to the method of citizen science the platform can raise awareness and capacity building (Wamsler and Raggers 2018) and with an increasing number of users and categories, the platform is under constant change. The current status of the ClimateScan has about 1000 registered users that (can) upload projects around the world. More than 60% of the users are younger than 34, and 51% of users are female. With over 5000 uploaded projects, the platform is considered to be the biggest inventory of 'blue-green' projects around the globe for international knowledge exchange. Currently, all the data points are categorized into 7 sub-groups (Water, People, Nature, Heat, Energy, Urban Agriculture and Air quality) holding over 20 categories, which are each assigned a different colour as shown in the legend to the left of the webpage (Fig. 3.5). Users of ClimateScan can create their own climate adaptation categories.

Most of the uploaded projects belong to categories related to NBS, SuDS, WSUD and BMPs that are designed to reduce the rate and quantity of surface water runoff from developed areas and to improve runoff water quality. Uploads on ClimateScan include: constructed wetlands, bio swales, green roofs and walls, permeable pavements, rainwater gardens, and floating structures on public and private property (Table 3.2). Along

Table 3.2 ClimateScan categories most used in The Netherlands with definitions

Category	Definition
Swale	A shallow vegetated channel designed to conduct, infiltrate and retain water. The vegetation filters particulate matter.
Constructed wetland	Wetland: Flooded area in which the water is shallow enough to enable the growth of bottom-rooted plants. Wetlands are constructed in urban areas to store water after stormwater events and improve water quality.
Green roofs	A roof with plants growing on its surface, which contributes to local biodiversity. The vegetated surface provides a degree of retention, attenuation and treatment of rainwater, and promotes evapotranspiration.
Floating urbanization	Floating or amphibious constructions as floating homes will adapt to variation of water levels (flooding, drought). Floating homes are constructed around the world to adapt to climate change.
Permeable pavement	A permeable surface that is paved and drains through voids between solid parts of the pavement. A permeable pavement is a surface that is formed of material that is itself impervious to water but, by virtue of voids formed through the surface, allows infiltration of water.
Hollow gully free roads	Roads that are constructed as drainage. An example is a surface flood pathway: Routes in which exceedance water flows are conveyed on the ground. Also referred to as 'hollow' or 'gully free' roads.
Sub-surface infiltration	A sub-surface structure into which storm water is conveyed, designed to promote infiltration to restore groundwater levels.
Heat stress measures	An upcoming category linked to implementation of green and blue measures in previous categories (swales, green roofs and walls, rain gardens, etc.) implemented to cool down the city and mitigate heat stress.

with uploading climate adaptation measures, problem areas are also mapped, where solutions can be implemented.

One of the Dutch medium size cities in The Netherlands that uses the ClimateScan platform is Groningen. Over 100 projects in the municipality of Groningen have been uploaded and categorized (see Fig. 3.6). Every city can use this platform to engage partners to upload climate adaptation measures and analyse their own situation as in Fig. 3.6.

ClimateScan is used in the international Young leaders programme from the GCA (based in Groningen and Rotterdam) for young professionals to get acquainted with global mitigation and adaptation projects.

ClimateScan allows analysing climate adaptation on a national level using the Dutch neighbourhood typologies as presented in Fig. 3.1 and

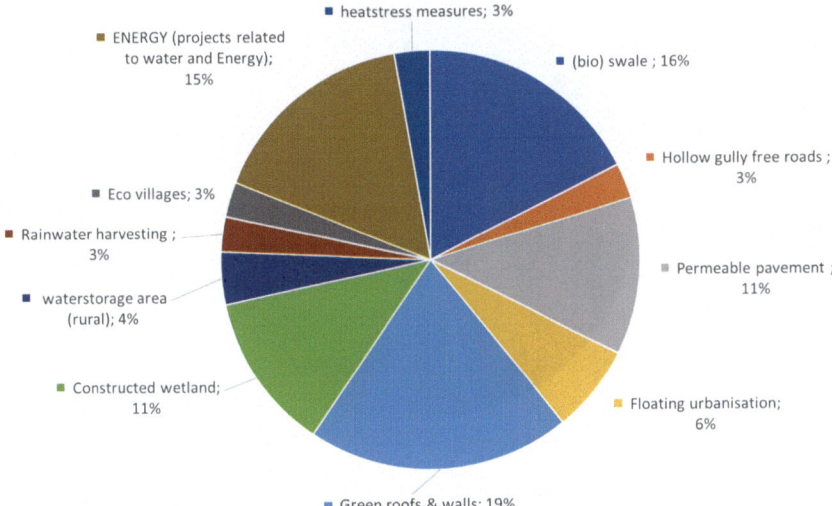

Fig. 3.6 Distribution of 100 projects in categories of the city Groningen, The Netherlands

Table 3.1. As an example, Fig. 3.7 illustrates that over 50% of the 168 locations of permeable pavement are located in the three more spacious and newer types of urban areas: villa district, community neighbourhood and suburbanization area. However, it shows as well that permeable pavement is also present in densely built up areas like working-class neighbourhood and historical centre. This information gives an evidence-based proof to urban planners that permeable pavement is implemented in any type of district but mostly is constructed in spacious districts built after 1970. Furthermore, examples of storm water infiltration projects in low-lying districts with high groundwater tables and low permeable soil in challenging can be found also in densely built up areas. This should inspire urban planners and storm water managers to design, plan and implement climate resilient measures with more confidence.

The ClimateScan platform is developed with support from several international projects serving the need of different stakeholders creating maps on climate resilience in different regions of the world. The platform

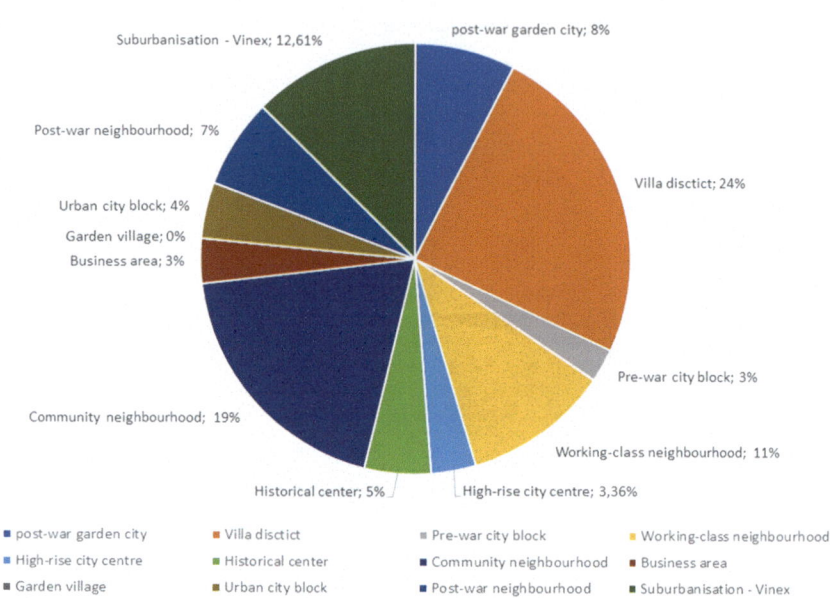

Fig. 3.7 Dutch neighbourhood typologies related to implemented Dutch permeable pavement

is applied in international city Climatecafés (Boogaard and Venvik 2019) and workshops in Semarang, Indonesia (Adi et al. 2020), and Johannesburg, Africa (Leal et al. 2020), targeting young professionals that were very helpful with uploading several projects around the world (Africa, Europe, Asia and South America), and the platform will be used in various new projects and ClimateCafés in the near future.

Climate Change Adaptation Platforms such as ClimateScan are an inspiration to stakeholders to make their cities more resilient to climate change. Challenges are reaching the public and stakeholders and keeping the content up to date. Potential topics for upgrades of these platforms have been identified with stakeholders. As mentioned, ClimateScan is a 'learning by doing' platform, and this comes with challenges. ClimateScan is depending on enthusiastic registered users uploading projects and categories. Some projects only have a location with a short description. The fact that anybody can upload projects (citizen science) makes quality control important. This is now done by volunteers that check and adjust new projects and categories out of their own interest.

3.5   REFLECTION ON FIVE PILLARS OF RESILIENCE

The discussed urban climate adaptation relies on threshold, coping and adaption capacity. The main path for reconstruction and retrofitting urban areas is aiming at **threshold capacity**. More and more Dutch municipalities have chosen to require that only above a certain amount of rainfall in a short timespan flooding of buildings is allowed. For example, no flooding of houses at 60 mm of rainfall within 1 hour (estimated return period of 50–100 years).

Next to this **coping capacity** is important. Flooding of streets is currently allowed and will stay allowed at a much higher frequency than the above threshold. In fact, in practice, municipalities accept flooded streets once every 1–2 years (20 mm–30 mm of rainfall within one hour). Note: for important streets the threshold will be higher and lower frequencies will be requested. It is expected that society has sufficient coping capacity to deal with those situations.

Lastly, **adaptive capacity** will be strengthened because people will learn from extreme weather conditions. After experiencing a local flooding or heat wave they might make more resilient choices in their professional and private lives. In an unexpected way, the approach of allowing water on the streets every 2 years might prove to be an effective way to reduce future damage to climate change. Finally, because streets are redesigned every 15–30 years, there are recurring possibilities to further adapt to future developments of weather extremes.

3.6   CONCLUSIONS

Because of the expected problems, Dutch municipalities have the task to take climate adaptation into account and to use all urban development projects and retrofitting or refurbishing of existing urban areas to step by step make the whole urban area climate proof. However, because there are so many other important requirements which need to be fulfilled when reconstructing a road or an urban area, climate resilience is often forgotten, or not fully implemented. It is argued that clear guidelines would help to meet the goals of climate resilient (re)design. Next to guidelines, two ways have been identified to encourage municipalities to choose for climate resilient projects. Both (a) the presentation of example designs for retrofitting streets with comparison of life cycle costs and benefits and (b) the access to real project examples seem to meet a need of local professionals. *The examples have been combined in a book which was widely requested and was reprinted twice.* The book is however not a guideline for how to act, but rather a book to show possibilities, after which municipalities still have to decide how they want to implement climate adaptation.

To show that climate resilient intervention has been implemented before, the platform ClimateScan with access to real project examples has been created. This and other platforms of examples can inspire different stakeholders that other designs than the 'standard solutions' are possible and can be more effective (larger storage volume, higher hydraulic capacity, etc.). Uploading projects by stakeholders themselves and sharing these examples show a high participation and will stimulate repetition of innovating solutions in the urban area to make cities resilient. Analysing the ClimateScan database, where solutions can be linked to the Dutch Neighbourhood typologies, allows to monitor the national progress of climate adaptation in The Netherlands and to identify opportunities for climate adaptation. The database gives an evidence-based proof to urban planners that over 2000 Dutch projects are implemented in any type of district. The results of this study will help urban planners and storm water managers with the designing, planning, climate adaptation measures with more confidence.

REFERENCES

Adi, H. P., Wahyudi, S. I., Boogaard, F., & Boer, E. (2020). Relevance of climate adaptation platforms in Asia: Pilot Central Java, Indonesia. *Journal of Water and Climate Change*, 2020 to be published.

Boogaard F., & Venvik G. (2019, June 12–14). *Knowledge exchange on climate adaptation best management practices for sustainable water management in resilient cities, international conference cities, rain and risk*, Malmo, Sweden.

Boogaard, F., Tipping, J., Muthanna, T., Duffy, A., Bendall, B., & Kluck, J. (2017, September 10–15). *Web-based international knowledge exchange tool on urban resilience and climate proofing cities: ClimateScan*. 14th IWA/IAHR international conference on urban drainage (ICUD), Prague.

Buck Consultants International. (2016, April 1). TEEB.stad: Kengetallen TEEB-stad tool (*The economics of ecosystems and biodiversity*). Nijmegen, 7 pp.

CAS. (2020). Website: www.climateadaptationservices.com/en/. Visited 12 May 2020.

Climate Ireland. (2019, October). *Visualization used at global climate change adaptation platforms as presented during 2nd international climate change adaptation platform workshop*, Dublin.

ClimateScan. (2020). Website: www.climatescan.org. Visited 12 May 2020.

Deltaprogramma Ruimtelijke adaptatie. (2014). *Deltabeslissing Ruimtelijke adaptatie: Het Deltaprogramma: een nieuwe aanpak*. De Haag: Ministerie van Infrastructuur en Milieu en Ministerie van Economische Zaken. www.rijksoverheid.nl/onderwerpen/deltaprogramma/inhoud/vijf-deltabeslissingen

GCA Global Commission on Adaptation. (2019, September 13). *Adapt now: A global call for leadership on climate resilience*. Groningen/Rotterdam: Global Center on Adaptation and Washington, DC: World Resources Institute. https://cdn.gca.org/assets/2019-09/GlobalCommission_Report_FINAL.pdf. Rotterdam.

IPCC Intergovernmental Panel on Climate Change. (2019). *IPCC special report on climate change, desertification, land degradation, sustainable land management, food security, and greenhouse gas fluxes in terrestrial ecosystems. Summary for policymakers*. Geneva: IPCC. https://www.ipcc.ch/report/srccl/. Visited 15 Aug 2019.

Kleerekoper L. (2016). *Urban climate design: Improving thermal comfort in Dutch neighbourhoods*. Doctoral dissertation, Delft University of Technology, Delft, The Netherlands, 378 pp. doi:https://doi.org/10.7480/abe.2016.11.

Klok, E. J., & Kluck, J. (2018). Reasons to adapt to urban heat (in the Netherlands). *Urban Climate, 23*, 342–351. Het klimaat past ook in uw straatje: De waarde van klimaatbestendig inrichten. Voorbeeldenboek.

Kluck, J., Boogaard, F. C., Goedbloed, D., & Claassen, M. (2015, November 3). *Storm water flooding Amsterdam, from a quick scan analyses to an action plan*. International waterweek 2015, Amsterdam.

Kluck, J., R. Loeve, W.J. Bakker, L. Kleerekoper, M.M. Rouvoet, R. Wentink, J.H. Viscaal, E.J. Klok en F.C. Boogaard (2017) Het klimaat past ook in uw straatje: De waarde van klimaatbestendig inrichten. Achtergronden. Hogeschool van Amsterdam, Faculteit Techniek, Onderzoeksprogramma Urban Technology, ISBN 978-94-92644-01-5, 49p.

Kluck, J., Loeve, R., Bakker, W., Kleerekoper, L., Rouvoet, M., Wentink, R., Viscaal, J., Klok, L., & Boogaard, F. (2018, April). The climate is right up your street, The value of retrofitting in residential streets, A book of examples. ISBN 978-94-92644-06-0, Amsterdam.

Leal Filho, W., Luetz, J. M., & Ayal, D. Y. 2020. Editors of Book: *Handbook of climate change management research, leadership, transformation*. Chapter:

Analysing climate change adaptation citizen science ClimateScan platform: Case South Africa. Springer (to be published in) 2020.

Majidi, A. N., Vojinovic, Z., Alves, A., Weesakul, S., Sanchez, A., Boogaard, F., & Kluck, J. (2019). Planning nature-based solutions for urban flood reduction and thermal comfort enhancement. *Sustainability, 11*(22), 6361. https://doi.org/10.3390/su11226361.

Stichting Rioned. (2015). *Leidraad Riolering D1100 (Sewer guide).* Ede: Rioned.

Tipping, J., Boogaard, F., Jaeger, R., Duffy, A., Klomp, T., & Manenschijn, M. (2015, November 3). *ClimateScan.nl: The development of a web-based map application to encourage knowledge-sharing of climate-proofing and urban resilient projects.* International waterweek 2015, Amsterdam.

Wamsler, C., & Raggers, S. (2018). Principles for supporting city-citizen communing for climate adaptation: From adaptation governance to sustainable transformation. *Environmental Science & Policy, 85,* 81–89.

CHAPTER 4

Flood Resilience of Critical Buildings: Assessment Methods and Tools

Manuela Escarameia and Andrew Tagg

Abstract Within the context of urban flood risk management, this chapter focuses on the vulnerability assessment and improvement of flood performance of buildings, in particular those that perform essential functions—designated as critical buildings. Three methods are presented as part of a framework for the assessment of building flood vulnerability and resilience. The first method (the "Quick Scan") is a "first pass", simple method to identify the assets most at risk of flood damage and easiest to tackle, leading to cost-effective interventions. The second method provides the necessary information and tools for selection and evaluation of flood resilient options for critical buildings while the third method (the Individual Building flood damage Tool, or IBT) allows detailed estimation of the extent of damage.

M. Escarameia (✉)
Engineering, HR Wallingford, Wallingford, Oxfordshire, UK
e-mail: m.escarameia@hrwallingford.com

A. Tagg
Floods and Water Management, HR Wallingford, Wallingford, Oxfordshire, UK

© The Author(s) 2021
R. de Graaf-van Dinther (ed.), *Climate Resilient Urban Areas*,
Palgrave Studies in Climate Resilient Societies,
https://doi.org/10.1007/978-3-030-57537-3_4

Keywords Flood resilience • Critical buildings • Critical urban infrastructure

4.1 Introduction

Due to the expansion of urban centres and the increased severity and frequency of wet weather events, it is now acknowledged that a range of methodologies and solutions for flood risk management are needed to protect people and assets from the impacts of floods. Within the portfolio of possible methods and tools, some are applied at the larger scales of a city, neighbourhood or street while others are applied at the scale of a single building. This chapter covers several assessment methods and mitigation solutions that were developed with the specific aim of understanding and improving the flood performance of buildings, and in particular those within an urban area that perform essential functions—designated as critical buildings.

The European project FloodProBE (www.floodprobe.eu), which focussed on flood protection of built environments and particularly on critical urban infrastructure, provides the main source of information for this chapter. Within the context of this project, critical urban infrastructure was defined as assets that are essential for the continuity of economic activities and for meeting the fundamental needs of the urban population. Critical urban infrastructure includes networks for transport (roads, railways), energy and water supply/sewerage, as well as information and communication services. This has been the subject of extensive research in recent years (e.g. Pearson et al. 2018; www.intact-wiki.eu; CIRIA 2010 and an industry-sponsored code of practice on the comprehensive assessment of the resilience of buildings, CIRIA 2019). But critical urban infrastructure includes also the so-called *"hotspot buildings"* (or *critical infrastructure buildings*), which provide an important role in protecting equipment and personnel associated with the above networks. They include hospitals and evacuation shelters, fire and police stations, electricity stations or substations, water and wastewater treatment plants, control centres of public transport and services, railway stations, and communication hubs.

Flooding of buildings can originate from a variety of sources, from rivers and the sea to groundwater, sewers and rainfall (pluvial). The damage that buildings experience is dependent on a number of variables, some of which are event-related (over-floor water depth, flow velocity, rate of water rise, presence of debris and contaminants, frequency and duration of inundation and timing); and some are building-related (type of structure and construction, materials used and their drying characteristics, services and their locations, as well as the age and condition of the building prior to being flooded).

Considerable research efforts have been made in trying to understand the effect of floods on buildings. Holistic approaches such as the Blue-Green city initiative pioneered in the UK or "sponge cities" are being applied to cities worldwide with the aim of recreating a water cycle as natural as possible to improve water management and amenity value (www.urbanfloodresilience.ac.uk; Chan et al. 2018). Earlier frameworks for urban flood resilience have been developed by Blanco and Schanze (2013), Kienzler et al. (2013), Ogunyoye and Dolman (2013), De Graaf (2012) while others have investigated and disseminated resilient technologies to prevent or limit ingress of flood water into buildings, for example in the SmarTest project, and others have advised on the possible role that building materials can have in mitigating damage and speeding up the recovery process (Maqsood et al. 2017; Bradley et al. 2016; Escarameia et al. 2012, CIRIA 2006).

4.2   RESEARCH ON RESILIENCE OF BUILDINGS

Much of the research carried out on buildings has been concerned with the development of flood-damage prediction methods, initially mostly for residential buildings but now much extended to non-residential properties. The Multi-Coloured Handbook (MCH 2019, online) is a guide to assessing the benefits of flood and coastal erosion risk management, to improve decision making on investments. This document includes methods for estimating damage to residential and non-residential property, and their contents, through depth/damage curves for England and Wales, information on protection measures and methods for estimating direct and indirect flood losses to schools, hospitals, utilities and transportation networks, to inform the prioritisation of actions on critical assets.

Walliman et al. (2013) presented a review of estimation methods in use in the UK, Germany, USA and Australia and suggested ways to improve on these methods in order to estimate damage to individual buildings, particularly non-residential ones. Blanco and Schanze (2013) suggested a framework based on building classification, using remote sensing data to establish roof surface as well as topological and building characteristics. From the selection of representative types of building, it is then possible to assess their vulnerability in terms of depth-damage functions which are defined based on the principal components of the buildings: floor height, building components and materials. The approach proposed was also tested on the city of Dresden, Germany (Blanco-Vogt et al. 2015). As noted by the authors, these methods were developed for the large-scale assessment of building vulnerability where data is scarce to allow general decisions at a neighbourhood/city level.

Golz (2016) reported on a process for quantifying and comparing the physical flood vulnerability of various building constructions, and evaluating and ranking the impact of alternative flood resilient building materials and constructions. More recently Balasbaneh et al. (2019) explored the feasibility of repairs to flood-damaged buildings, taking into account not only building costs but also sustainability aspects such as the CO_2 emissions after repairs in a flood zone, in non-flood situations and when a flood hits the building. From the five types of building material studied (brick, concrete block, steel framework, timber and precast concrete framing), although timber was the best choice for constructing the building where flood risk is small, in case of increased flood occurrence, precast concrete framing construction indicated better performance by releasing less CO_2 after the repair stage. Cost analyses (full life cycle assessment (LCA) that includes global warming potential and life cycle cost (LCC)) indicated that timber and steel frame were the worst options in a high-risk flood zone, while brickwork was the most sustainable one. Using timber as a building material in flood zones is therefore not recommended as alternative materials such as brick have better performance in terms of both LCC and global warming mitigation.

From the point of view of flood vulnerability and resilience of critical buildings, it is apparent that a much higher level of detail is required than what is applied to domestic dwellings. One of the reasons is that these buildings, unlike domestic constructions, often include several different types of construction within the same premises and characterisation according to construction type is insufficient (Walliman et al. 2013). Walliman et al. (2013) reported on an assessment methodology that identifies the likely level of damage to individual buildings and described a tool (the Individual Building flood damage Tool—IBT) for estimating in detail the extent of damage where the results are expressed as the predicted costs of remedial works. An advantage of using this tool is that plans can be developed in advance of flood events to ensure prompt rehabilitation of buildings, which are particularly relevant for critical infrastructure buildings.

4.3   Methods for Assessment of Flood Resilience of Critical Buildings

Although this chapter introduces three specific tools or approaches for assessing the vulnerability and resilience of critical buildings, it is important to recognise that these should sit within an overarching risk assessment framework, considering the hazard, the consequences and their mitigation. Such a risk framework is shown in Fig. 4.1, developed in the EU-funded INTACT project, which considered the impact of extreme weather on critical infrastructure. It is noted that the risk framework is not linear in that it does not terminate at the risk control stage but is a process where the risk is analysed, assessed and managed on a cyclical basis.

The risk process in INTACT was based on the IEC 60300 procedures (now superseded by IEC/ISO 2019). As such it comprises a widely used assessment approach, which is supported by a range of tools that can be

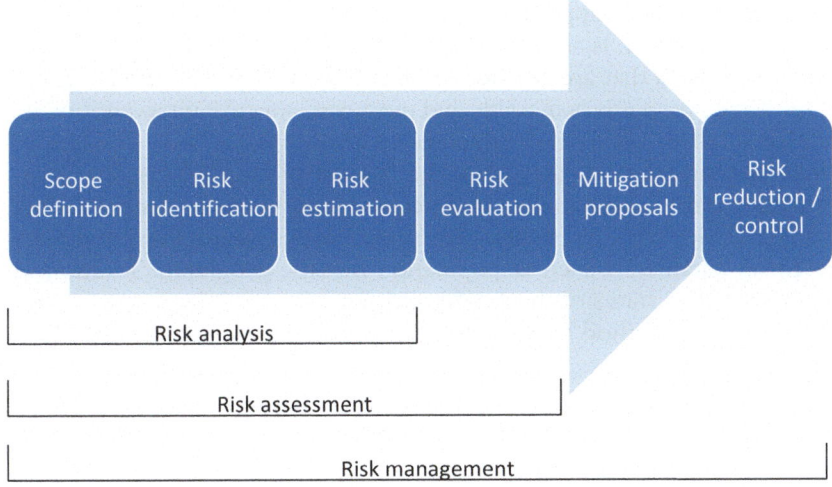

Fig. 4.1 Layout of INTACT risk management process. (Adapted from www. intact.wiki.eu)

used at each of the six assessment steps. So the building tools here do not work in isolation from general risk approaches. For example, there are a range of methods for choosing the "optimum" mitigation measure, which includes Cost Benefit Analysis, multi-criteria methods, indices and ranking.

Three different methods are presented next as part of a framework for the assessment of building flood vulnerability and identification of improvement measures targeted at critical infrastructure. The first method, called the "Quick Scan", is, as the name implies, a "first pass" that allows identification of the assets most at risk of damage and easiest to tackle, leading to cost-effective interventions. The second method is concerned with providing the necessary information and tools for selection and evaluation of flood resilience and resistant options which can be used in critical buildings. The third method, or tool to be more precise, is the Individual Building flood damage Tool (or IBT for short) for the detailed estimation of extent of damage.

4.3.1   Method 1—The "Quick Scan"

In cities, with all their associated complexity, the number and type of flood vulnerable infrastructure is likely to be large and very varied. To facilitate action prioritisation, the Quick Scan method was developed (Zevenbergen et al. 2014, where application of the method to neighbourhoods in Bangkok, Thailand, is also described). This method enables quick identification of assets that are at most risk from flooding so that effective, but at the same time easily achievable and most cost-effective interventions are implemented to alleviate the damage to these assets. These are termed "low-hanging fruit" as they offer the "easiest picking". This very pragmatic approach is best applied to a neighbourhood or city district (rather than to whole cities due to their complexity) with a well-defined boundary based on parameters such as population density, geographic location or socio-economic status of the neighbourhood.

This simple method uses the following five steps to identify the "low-hanging fruit":

Step 1—Identification of critical infrastructure assets and ranking of criticality

Firstly, the critical urban assets, including both networks and buildings, and the relationship between them (i.e. dependence of assets on one another) are identified. A network analysis is then carried out by describing the critical networks, studying the effects of failure of one element on the performance of the network, the effect of failure of the network on other networks and ultimately the effect of failure of one or more networks on the urban community.

Step 2—Analysis of vulnerability, that is, the exposure and sensitivity of the critical assets

In this step it is useful to define thresholds to help define the sensitivity of elements (e.g. a clear threshold is defined in Dordrecht, The Netherlands, for transformer stations which will fail if flood water depths exceed 0.3 m). Although clear thresholds are not always available, they will allow analysis of potential flood parameters and probabilities ("the exposure") to be carried out. Being the most readily available parameter, the main flood parameter is water depth, but flood duration and velocity may also be used as parameters for vulnerability. Flood exposure analysis is normally based on historical data and/or model simulation of potential floods and the sensitivity depends on the flood resistance and recoverability of the critical assets. Flood resistant assets are those that can withstand a particular flood water depth without damage or failure while flood resilient assets experience no permanent damage, retain structural integrity and can resume normal operation soon after the flood has receded. Vulnerability can be expressed in monetary terms or be based on indicators such as the duration of outage, the number of people affected or combinations of these.

Step 3—Determination of the severity of the impact

This step involves the assessment of: the effect of failure of the critical assets (nodes and connections) on the delivery of service (first order); the effect of failure of a network (or part of one) or node on other networks; and the likelihood of failure (flood exposure and sensitivity).

Step 4—Option identification to mitigate the effects of flooding and their associated costs

The options comprise flood proofing (resilient and resistant) construction and retrofitting techniques ranging from simple interventions such as temporary closures to permanent elevation of a building or road. Another way of alleviating the impacts from failure is by reducing the criticality of the sensitive elements. This can be achieved by introducing redundancy in the network so that if one node fails, the network will still function.

Step 5—Identification of the "Low-hanging fruit"

The previous steps lead to the identification of assets where actions can be undertaken at the lowest cost but with high impact. This can be taken as part of a wider range of interventions to fully protect the urban critical infrastructure.

4.3.2   Selection and Evaluation Tools for Flood Proofing of Buildings

A number of researchers have worked on and suggested a range of flood proofing measures for general buildings and these have been published in guidelines in various countries, with a variable degree of detail (see for example Vassipoulos et al. 2007 or Escarameia et al. 2012). However, specific guidance for critical buildings and retrofit is less common, and so are the evaluation methods to assess the effectiveness of these measures. This section describes a design tool that was developed to allow policy and decision makers and designers to narrow down the range of options for flood proofing of critical buildings (FloodProBE 2012a).

Flood proofing can have different meanings in different countries (and for different researchers) but it is defined in FloodProBE (2012a) as measures that allow buildings to cope well with the impact of floods and comprise a number of methods, as listed below (see also Escarameia and Stone 2013):

- Wet flood proofing or wet proof construction allows temporary flooding of the lower parts of a building, the building materials being easily repaired or replaced
- Dry flood proofing or dry proof construction prevents water ingress by the use of water resistant materials and/or coatings; in addition building materials should be able to dry easily
- Elevating the building, in stilts or mounds, and ensuring the connection to infrastructure (such as access roads) is also flood proof
- Floating buildings, where the building is founded on a floating structure that is permanently located in the water
- Amphibious buildings, which are located on the ground with a traditional foundation combined with a floating foundation and will float during flood events

- Temporary flood barriers, which are only placed when a flood is expected to damage the building and removed completely when levels have receded
- Demountable flood barriers, which are partly preinstalled and require full deployment before a flood event
- Permanent food barriers, specifically constructed to protect a building or group of buildings, for example, a dike around a hotspot building, an integrated flood defence system protecting a whole area of critical buildings.

For the particular case of critical buildings (hotspots), recommendations were also made on the applicability of the various concepts, as illustrated in Fig. 4.2 with regard to the dependence of flood proofing methods on flood depth and duration.

The interactive flood proofing design tool (in Excel spreadsheet format) helps in the choice of the most appropriate flood proofing concepts for buildings at different stages of the urban development process and

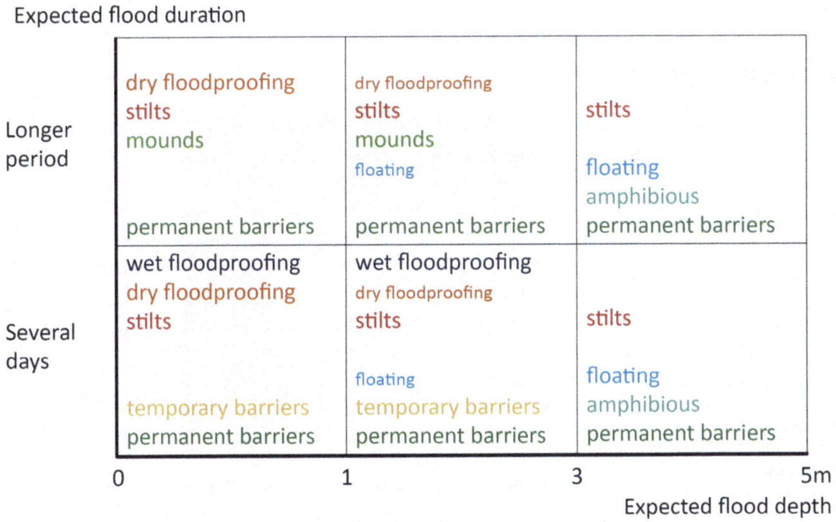

Expected flood duration

Longer period	dry floodproofing stilts mounds permanent barriers	dry floodproofing stilts mounds floating permanent barriers	stilts floating amphibious permanent barriers
Several days	wet floodproofing dry floodproofing stilts temporary barriers permanent barriers	wet floodproofing dry floodproofing stilts floating temporary barriers permanent barriers	stilts floating amphibious permanent barriers

0   1   3   5m

Expected flood depth

Fig. 4.2 Applicability of flood proofing measures according to flood level and duration (FloodProBE 2012a)

consists of three stages: (1) the Relevance Map, (2) the Selection Tool and (3) the Evaluation Tool.

The Relevance Map provides a first check to evaluate how relevant the application of flood proof measures is. The relevance is assessed based on two parameters: the service area (i.e. how big an area/how many people rely on the service provided by the critical building) and the magnitude of the flood event (i.e. how many people will be affected)—see Fig. 4.3. For example, hotspots such as airports and food distribution centres are generally large-scale facilities that serve major populations. If they are flooded, they impact greatly on people and the economy, whereas a district bus station would have a much smaller impact. A Relevance Map thus gives an indication of the relative importance of flood proofing a particular hotspot building based on flood impacts and the service area of the building.

The Selection Tool, in the next stage of the process, when possible measures for flood proofing are being considered, helps identify the most feasible flood proofing concepts based on information on location and building characteristics. This tool, by working at a qualitative level, still requires a limited amount of information, but more than in the first phase.

The Evaluation Tool is used in the final, decision making phase. Finding the optimum solution (both from a technical and cost viewpoint) is a

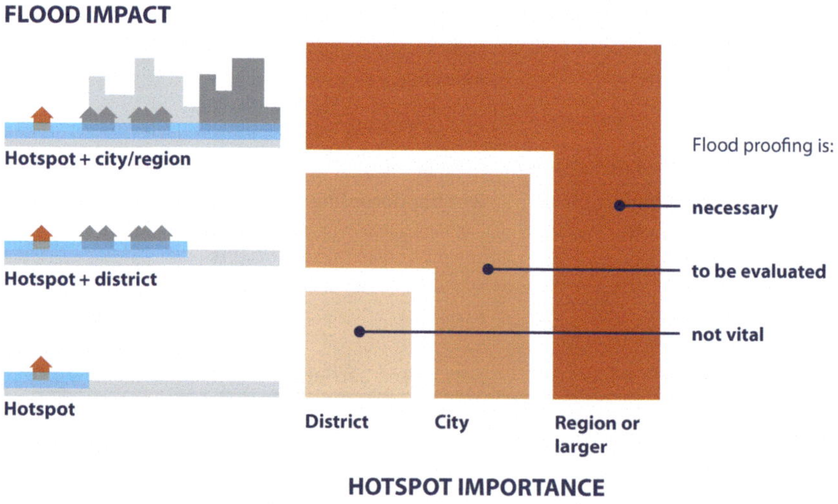

FLOOD IMPACT

Fig. 4.3   Illustrative Relevance Map (FloodProBE 2012a)

complicated process because various parameters play a role, related both to the properties of the building (site area/perimeter, building area/perimeter, land cost, height, service area) and the type and characteristics of the flood hazard (depth, frequency, flood onset time from time of warnings). The tool assists in this process through the use of a database of reference flood proofing products that was built from various sources: research publications, data from governmental agencies (e.g. UK Environment Agency and the US FEMA) as well as data from product suppliers. The Evaluation Tool provides detailed, quantitative information on the costs of several possible options for flood proofing a specific hotspot. Relatively detailed information on the hotspot (including whether it is a retrofit of an existing building and whether the site is protected by levees), flood characteristics and location characteristics need therefore be available for application of this tool. Figure 4.4 illustrates one of the most relevant components of this tool, the space required for the different measures—others include the installation time, the height range and cost.

Validation of this model was carried out through three case studies (a hospital in the USA, a bank headquarters in Italy, an electricity substation in the UK) and a pilot study (St Francis Hospital, The Netherlands) subjected to different types of flood (coastal, riverine and pluvial) and having

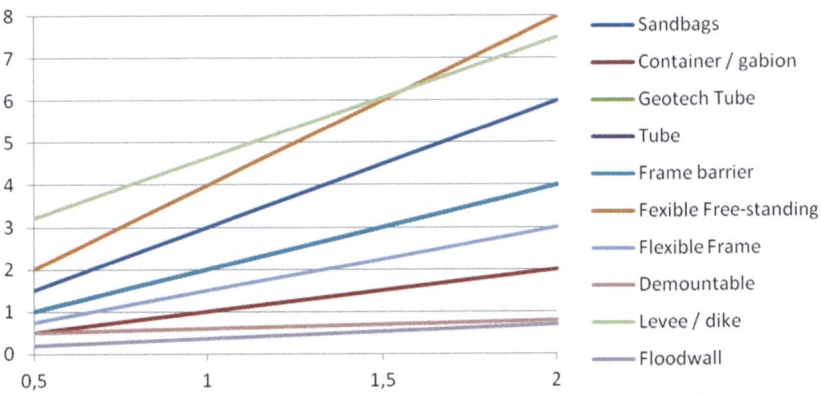

Fig. 4.4 Example of required distance from the boundary of the hotspot (y-axis) as a function of flood height (x-axis) required for different flood protection products, in metres (FloodProBE 2012a)

different building types. For each case study the flood event was defined in terms of frequency, depth, extent and rate of rise of the flood.

4.3.3   The Individual Building Flood Damage Tool (IBT)

The third of the methods covered in this chapter takes us into the higher level of detail that is required when determining and mitigating flood damage at building level. Due to the variety of designs and constructions in these buildings, categorising them into classes with regard to their vulnerability to flooding is not the right approach. Instead, predicting the effects of flooding and costs of reinstatement of these buildings requires estimation of damage at an elemental level (e.g. wall, apertures, floor, services) rather than considering the building as a whole. The method is an assessment methodology that distinguishes between buildings built using different construction methods and materials (FloodProBE 2012b). It helps identify key risks in order for buildings to be retrofitted to improve their robustness against flooding, facilitates planning of rehabilitation response and helps target investment for reducing future vulnerability. Although particularly relevant for critical buildings, it can be applied to other types.

The methodology shown in Fig. 4.5 was the basis of a damage prediction tool (Individual Building flood damage estimation Tool or IBT), applicable mainly in the European context. This is a simple-to-use tool, designed to be used by building professionals throughout Europe, which does not assume previous specialist flood knowledge but enables them to predict the cost of flood damage to individual buildings depending on the nature of the flood event and the individual characteristics of a particular building. Although the cost data was based on UK prices in 2012, there are integral conversion factors that allow adjustments to be made for the different building costs in the EU countries. Prices and conversion factors can be updated regularly to reflect changes over time.

It is noted that before calculating the damage to a building, it is worth checking that the building is structurally stable against floods as otherwise the damage calculations would be useless. The likelihood of collapse can be determined using depth/collapse curves as well as by structural calculations.

The tool is based on a spreadsheet with a simple user interface which requires the user to first insert the following basic information which relates to the context:

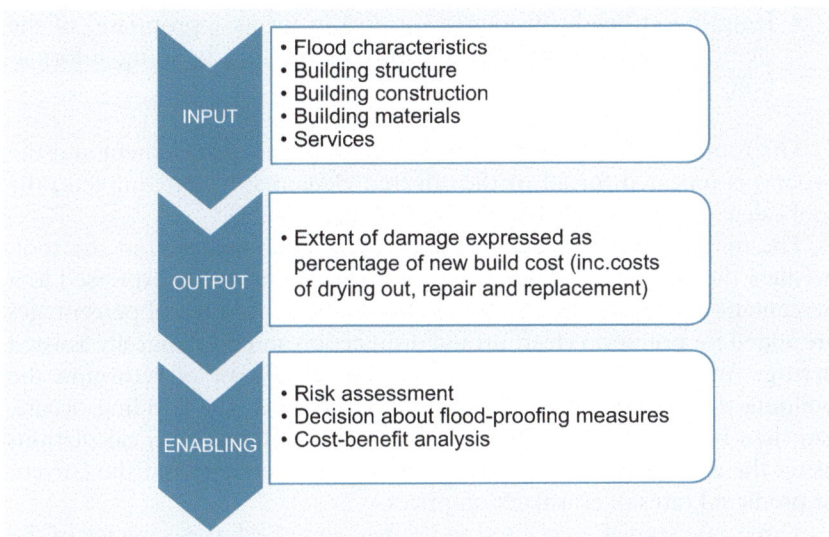

Fig. 4.5 Individual building damage methodology and tool. (Adapted from FloodProBE 2012b)

- Name of the premises
- Flood characteristics (flood depth (m), flood duration (days), pollution, velocity/debris)
- Currency (£ or €)
- Region (EU country). This then automatically calculates a regional adjustment (accounting for differences in building costs in various EU countries).

Further information needs to be input concerning the building itself:

- Building element (external walls, internal walls, floors, windows, external doors, stairs, services and finishes). These are selected in turn, and the following data for each is inserted:
- Type (e.g. choice for external walls is masonry monolithic, cavity wall, concrete, steel, curtain wall, timber)
- Description

- Length and height in metres (or area in m² as appropriate) of the element (this needs to be measured from the building information/plans).

The tool then calculates the repair cost of the chosen element and the process is repeated for all of the affected elements; when complete, the tool calculates the overall cost of the building repairs.

The initial output, based on the database contained within the tool, predicts the cost of cleaning, repair or replacement and is expressed as a percentage of the new-build cost of each element. Additional percentages are added for pollution clean-up and disinfection and mechanically assisted drying. An approximate indication of the actual cost of returning the building to use, depending on where it is and when the flooding occurs, can then be produced by the tool when it is combined with calculations using the areas, lengths or numbers of affected elements and the current or predicted rates of construction prices.

Three case studies were used to validate and check the accuracy of the predictions made by the tool. These were selected from data supplied by AXA Insurance on three premises that were flooded in 2007 in the Sheffield area of the UK. In each case, the data included a description of the flood event, an outline specification and photographs of the building, the area of the floor plan, and the cost of remediation. Due to the lack of detailed construction drawings, some assumptions had to be made about the exact details of the building construction and areas of elements. The comparison of the tool's estimate of costs of repair against the actual costs of the case study buildings resulted in a close match in two cases (an engineering workshop and a wine store and warehouse—88% and 95% accuracy respectively), while the estimate for the third case study, an office building, was only 63% accurate. It was concluded that this was due to insufficiently detailed building data (particularly about high levels of finishes) and damage cost data on individual constructions and materials. This indicated the need for further detail and refinement of the latter, and that obtaining detailed information from real flooded critical buildings can offer some challenges due in part to commercial or security sensitivities. The tool was therefore regarded as a demonstration of the methodology for calculation of flood damage costs based on flood characteristics, detailed constructional information and associated damage factors. It was expected that it would serve as a springboard for the development of a fully functional tool that can be used for buildings of all types throughout

Europe. It is noted that when assessing feasibility and viable flood proofing options for critical buildings, conventional cost-benefit analysis concentrating on buildings alone is not the most appropriate tool. Indirect impacts that reflect the wider interdependencies of critical buildings and associated networks are important to consider but the calculation of indirect impacts was not part of this tool.

4.4 Conclusion

This chapter introduced some methods to improve the coping and recovery capacities of urban centres when impacted by floods, by concentrating on critical buildings as these can be the centres around which a city will adapt to increasingly uncertain climate conditions. Such buildings may also exhibit high social vulnerabilities because of the important services that they provide. Although these methods provide answers to a number of resilience and vulnerability issues, other relevant questions emerge. For example, in relation to resource usage or associated greenhouse gas emissions, how sustainable are flood repairs that rely on replacement of light weight resilient materials compared with using heavy duty materials in construction? What can we do to minimise material usage taking into account whole life costs? These are areas that merit further and continued research, and ultimately enable a transformative pro-active climate resilient society.

References

Balasbaneh, A. T., Bin Marsono, A. K., & Gohari, A. (2019). Sustainable materials selection based on flood damage assessment for a building using LCA and LCC. *Journal of Cleaner Production, 24*, 844–855.

Blanco, A., & Schanze, J. (2013). Conceptual and methodological frameworks for large scale and high resolution analysis of the physical flood vulnerability of buildings. In F. Klijn & T. Schweckendiek (Eds.), *Comprehensive flood risk management* (pp. 591–598). London. ISNBN 978-0-415-62144-1: Taylor & Francis Group.

Blanco-Vogt, A., Haala, N., & Schanze, J. (2015). Building parameters extraction from remote-sensing data and GIS analysis for the derivation of a building taxonomy of settlements – A contribution to flood building susceptibility assessment. *International Journal of Image and Data Fusion, 6*(1), 22–41.

Bradley, A. C., Chang, W. S., & Harris, R. (2016). The effect of simulated flooding on the structural performance of light frame timber shear walls – An experimental approach. *Engineering Structures, 106,* 288–298.

Chan, F. K. S., et al. (2018). "Sponge City" in China—A breakthrough of planning and flood risk management in the urban context. *Land Use Policy, 76,* 772–778.

CIRIA. (2006, July). Report no. WP5C Final Report – Laboratory tests, by M. Escarameia, A. Karanxha and A. Tagg.

CIRIA (2010). *Flood resilience and resistance for critical infrastructure* (C688). CIRIA.

CIRIA (2019). *Code of practice for property flood resilience.* C790. London.

De Graaf, R. E. (2012). *Adaptive urban development. A symbiosis between cities on land and water in the 21st century.* Inaugural lecture. Rotterdam University Press.

Escarameia, M., & Stone, K. (2013). *Technologies for flood protection of the built environment.* Guidance based on findings from the EU-funded project FloodProBE. Report number WP05-01-13-03. www.floodprobe.net

Escarameia, M., Tagg, A., Walliman, N., et al. (2012). *The role of building materials in improved flood resilience and routes for implementation* (pp. 1303–1309). Flood risk 2012 the 2nd European conference on flood risk management. Science, policy and practice: Closing the gap, Rotterdam. ISNBN 978-0-415-62144-1.

FloodProBE. (2012a). *Construction technologies for flood proof buildings and infrastructures; Technologies for flood-proofing hotspot buildings.* Deliverable Report D4.3, Report number WP4-01-12-11. www.floodprobe.eu

FloodProBE. (2012b). *Assessment of the vulnerability of critical infrastructure buildings to floods.* Deliverable Report D2.2, Report number WP02-01-12-05. www.floodprobe.eu

Golz, S. (2016). *Resilience in the built environment: How to evaluate the impacts of flood resilient building technologies?* E3S web of conferences. https://doi.org/10.1051/e3sconf/2016

IEC/ISO. (2019). Risk management. *Risk assessment techniques.* IEC 31010, Edition 2.0, 2019–06.

Kienzler, S., Pech, I., Kreibich, H., et al. (2013). Coping with floods: Preparedness, response and recovery of flood-affected residents in Germany after 2002. In *Proceedings of international conference on flood resilience: Experiences in Asia and Europe.* Exeter.

Maqsood, T., Wehner, M., Edwards, M., Ingham, S., & He, D. (2017). *Testing of simulated flood effect on the strength of selected building components.* Bushfire and Natural Hazards CRC.

MCH. (2019). Online. https://www.mcm-online.co.uk/handbook/. Accessed Dec 2019.

Ogunyoye, F., & Dolman, N. (2013). Thinking more broadly about flood resilience. In *Proceedings of international conference on flood resilience: Experiences in Asia and Europe*. Exeter.

Pearson, J., Punzo, G., Mayfield, M., Brighty, G., Parsons, A., Collins, P., Jeavons, S., & Tagg, A. (2018). *Environment Systems and Decisions, 38*(3), 318–329, Springer US.

Vassipoulos, A., Ashley, R., Zevenbergen, C., Pasche, S., & Garvin, S. (Eds.). (2007). *Advances in urban flood management*. Rotterdam: Balkema.

Walliman, N., Baiche, B., Ogden, R., et al. (2013). Estimation of repair costs of individual non-domestic buildings damaged by floods. *International Journal of Safety and Security Engineering, 3*(4), 290–306.

www.floodprobe.eu

www.floodresilience.eu

www.intact-wiki.eu

www.urbanfloodresilience.ac.uk

Zevenbergen, C., Kolaka, K., van Herk, S., et al. (2014). Assessing quick-wins to protect critical urban infrastructure from floods: A case study (three urban communities) in Bangkok, Thailand. *Journal of Flood Risk Management*.

CHAPTER 5

Recovery Capacity: To Build Back Better

Frans H. M. van de Ven, Fransje Hooimeijer,
and Piet Storm

Abstract The ambition to Build Back Better after a serious flood disaster is a complex challenge. A comprehensive, multidisciplinary redevelopment planning process is required to reduce the flood risk and meanwhile create sustainable solutions that bring added value to society every day. General planning principles can be formulated on how to develop the physical conditions for flood resilience, while building a better place to live and work. Scoping and the charrette method are to be applied for pairing and

F. H. M. van de Ven (✉)
Urban Water Management, Delft University of Technology,
Delft, The Netherlands

Urban Land & Water Management, Deltares, Delft, The Netherlands
e-mail: frans.vandeven@deltares.nl

F. Hooimeijer
Environmental Design and Technology, Delft University of Technology,
Delft, The Netherlands

P. Storm
Water Management Department, Delft University of Technology,
Delft, The Netherlands

© The Author(s) 2021
R. de Graaf-van Dinther (ed.), *Climate Resilient Urban Areas*,
Palgrave Studies in Climate Resilient Societies,
https://doi.org/10.1007/978-3-030-57537-3_5

85

integrating disciplinary results, to co-create better plans in an interdisciplinary planning process. Two disaster recovery cases in Japan, after the 2011 Tohoku tsunami, and one case on Grand Bahama, after the 2019 Hurricane Dorian, were studied by multidisciplinary teams of students and staff to investigate how far Building Back Better was, or is to be, realized. This was done by confronting the practice of the reconstruction process and the resulting plans with the guiding principles for the physical concepts and interdisciplinary planning approach. Practice shows that Building Back Better is suffering from a lack of integration of disciplinary solutions, guided by existing planning regulations and practices and driven by the need for flood safety and by the urgency of reconstruction works.

Keywords Recovery capacity • Build Back Better • Multidisciplinary approach • Spatial planning • Guiding principles • Flood risk reduction

5.1   INTRODUCTION

The need for disaster risk reduction and improved disaster management stands beyond doubt. Disaster response is often well organized, as part of disaster preparedness. This phase of emergency relief accompanies the phase of recovery and reconstruction. Survivors want to build up their regular life again, but in a way that such a disaster 'can never happen again'. They want to Build Back Better and strengthen the protection level of their living environment. Rarely the reconstruction process is seen as an opportunity for fundamental changes in the social and economic development of a region. More often, the first step is to rebuild what was lost.

The United Nations Office for Disaster Risk Reduction (UNDRR) identifies Prevention, Preparedness, Response and Recovery as key components of risk reduction; however, recovery is the final and often least developed part of this framework (UNISDR 2005). That is why it was made one of the four priority areas of the Sendai Framework for Disaster Risk Reduction: Enhancing disaster preparedness [...] to "Build Back Better" in recovery, rehabilitation and reconstruction (UNISDR 2015).

This chapter will concentrate on the recovery from tsunami- and hurricane-hit areas. The recovery in Japan from the Tohoku tsunami in 2011 and on the Bahamas from Hurricane Dorian in 2019 were investigated by multidisciplinary teams of students and staff from TU Delft, to see how they Build Back Better. In Japan the reconstruction activities were ongoing when the cases were studied, while in the Bahamas, a few months after Dorian hit the islands, recovery planning was in its inception phase. The data on which this chapter is based are collected in close collaboration with Waseda University, Tokyo, Tohoku University, University of the Bahamas, the local authorities and many others. The teams visited the sites; on-site workshops with the local stakeholders were very important for appropriate understanding of the cases. The students in the interdisciplinary project groups of Yuriage (Areso Rossi et al. 2018; Vafa 2018; Möhring 2018; Mustaqim 2018; Dobbelsteen 2018, van Dijk 2018; Glasbergen 2018) and Otsuchi (Roubos 2019; Salet 2019; Filipouskaya 2019; Nederlof 2019; Yasaku 2019; Mujumdar 2019; Rao 2019; Broere et al. 2019; Blom et al. 2019; Van den Berg et al. 2019; Höller and van de Wiel 2019; Li et al. 2019; Prida Guillén 2019) used all the information for making their theses. The research and education project of Grand Bahama is still ongoing. Other experiences with the recovery process were collected in New Orleans after Hurricane Katrina and in Houston/Galveston after Hurricane Harvey.

Building Back Better (UNISDR 2017) is a complex process, not only because of broken infrastructure but also due to a lack of a well-functioning organization. Reconstruction authorities have to deal with the trauma of survivors on the one hand and the requirements of investors, and regional, national and international authorities on the other. The only way to Build Back Better is to organize a comprehensive, multidisciplinary planning and redevelopment process. This process will be highly situation- and site-specific. General planning principles can be formulated on how to develop the physical conditions for flood resilience, while building a better place to live in and work. These principles will be formulated and used to reflect on cases in Japan and the Bahamas, to give general recommendations on how to Build Back Better.

5.2   Guiding Principles

Guiding principles for the (re)construction of a flood- and climate-resilient urban environment can be split into principles on the physical concepts and principles in the planning process.

5.2.1   Physical Concepts

Safety first is a logical starting point for every reconstruction plan. The first question to ask is: Should we (re)build in these flood-endangered areas and, if so, how do we rebuild in a safe way? Space is to be made available for infrastructure needed to create safety, while extra protection is to be provided for critical infrastructural objects and networks as well as for the most vulnerable groups of the population.

A **risk approach** is necessary to Build Back Better. Risk is defined as the product of the probability of exposure[1] to a certain hazard and the damage sensitivity to this exposure. The **vulnerability** of our society includes a lack of capacity to adapt to newly emerging risks. Hazard reduction, hence, is a first, but often difficult, way of reducing risk. Measures to reduce exposure also provide effective ways of reducing risk. Damage sensitivity can often be diminished by the smart design of critical infrastructure and buildings.

Only **four ways** to reduce the risk of flooding are available (Van de Ven et al. 2009): the first is improvement of flood protection and drainage system; the second is to change the topography—digging a hole creates storage capacity and the excavated soil can be used to construct a mound for flood defence; the third is to make the construction of buildings and infrastructure more water-robust; and the fourth is to improve people's level of preparedness.

Four **capacities** are to be strengthened to reduce vulnerability (De Graaf et al. 2009). Threshold capacity is installed to protect against damage up to a certain level (e.g. levee, pump). If this is exceeded, coping capacity is needed to reduce the damage. Recovery capacity is required to minimize damage of the slow reconstruction process. Last but not least, adaptive capacity is needed to modify the system to unexpected changes in conditions that the site and society are exposed to.

The **three-point approach** (Fratini et al. 2012) to design is an important concept for Building Back Better. The first point in Fig. 5.1 represents a protection facility, designed to provide protection for extreme conditions that occur only once in so many years, the so-called design return period. Point two represents a situation where this protection level is exceeded (i.e. conditions are even more extreme and the protection

---

[1] For definitions of **Exposure, Hazard, Sensitivity** and **Vulnerability**, see (IPCC 2007, 881 and 883) and (IPCC 2014: Annex II: Glossary).

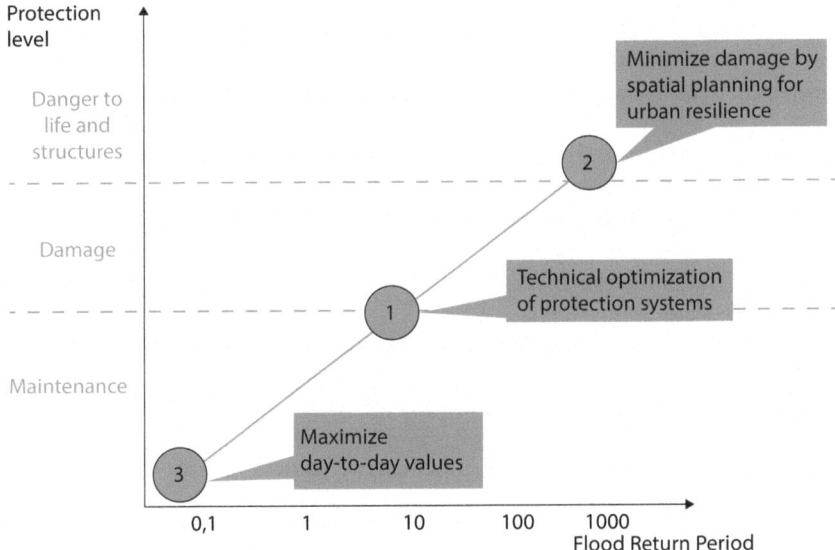

Fig. 5.1 Three-point approach for flood risk management. (Modified after Fratini et al. 2012)

system fails). This point emphasizes the need for a design aimed at minimizing the damage of that failing system by maximizing its coping and recovery capacity. The third point represents the everyday situation. Instead of being a hindrance, the facilities ought to provide added value and/or services to society every day. Multifunctional protection measures are essential.

Hybrid solutions: a balanced combination of traditional grey solutions, smart and nature-based solutions helps create a more resilient urban environment. Hybridity is not only a technical issue, it also refers to the need for spatial and governance hybridity (Sugano and Lu 2019). Spatial hybridity is about visibility, legibility and connectivity of blue-green elements in the urban landscape, while governance hybridity handles the balance between public, collective solutions versus solutions at the private, individual level.

Hybridity is also relevant in socio-economic reconstruction, balancing reconstruction of homes and economic activity. Lasting employment is a prerequisite for successful recovery of the area.

5.2.2   *Interdisciplinary Planning Approach*

Comprehensive planning of reconstruction requires an interdisciplinary approach. Many involved disciplines in urban development need to connect to their part of the problem. This can be done with **scoping** (Hooimeijer et al. 2018), carried out according to the **charrette method** (Lennertz and Lutzenhiser 2014).

The **charrette method** suggests a series of steps where disciplines are twinned in subgroup discussions and the size of each subgroup is gradually increased until the final session, when one large group discussion is held with all disciplines (Fig. 5.2).

The first stage of the process consists of the **scoping** method. This is the monodisciplinary analysis in which concepts and/or measure sets are selected and evaluated. What is important is that the different disciplines use comparable values to be able to explain and merge their chosen concepts with ones from other disciplinary perspectives. In the projects presented in this chapter the '4P approach' by Duijvestein and Van Dorst (2006) was used.

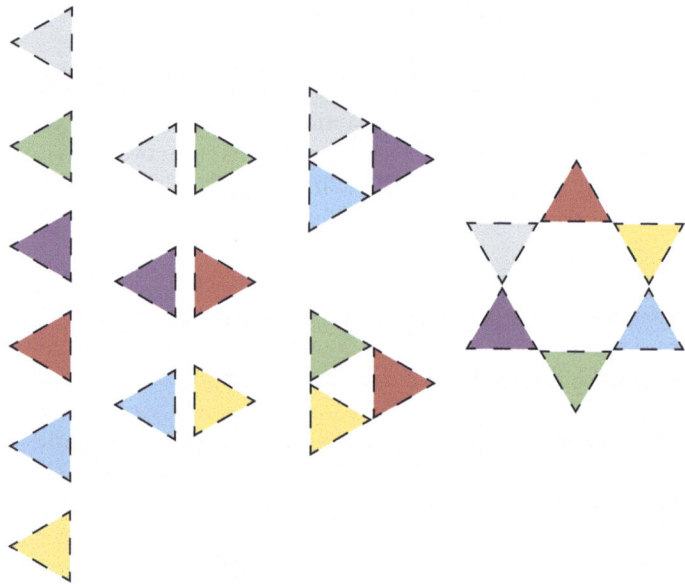

Fig. 5.2 Schematic representation of the charette approach

The next step consists of pairing up disciplines to discuss their scope and merge them. The value system is critical in the communication in order to organise in-depth discussions. The last two steps comprise of discussion in a group of three and finally, again, with all the disciplines present. Each step creates new combined scopes with integrated measure sets or concepts. This co-creative approach can be applied at different spatial scales, to confront large-scale planning ideas with small-scale, much more detailed plans to test the feasibility of the ideas.

This approach creates a 'professionals plan' in which all possible measures from the different disciplines are brought together through a synergy. The most desired variants can then be presented to local residents in co-creative sessions, so they can participate in the propositions to tune them towards local needs.

Balancing the role of local residents, representatives and politicians in the planning process with the power of external experts from different disciplines and representatives for regional and national governments and funding agencies is a delicate task. Seizing the opportunity, brought by the disaster, to Build Back Better is understood differently by the various parties. Often, a Redevelopment Authority is created to organize the recovery process; they have the hard role of directing the reconstruction process in order to make a future-proof plan without compromising the interests of other involved parties.

5.3   Cases

5.3.1   Japan

On March 11, 2011, Japan experienced a magnitude 9 earthquake (Fig. 5.2) that caused an enormous tsunami felt across the Pacific Ocean. Waves with heights of up to 40 m destroyed most of the eastern coastline in the Tohoku region; 560 km² of land were inundated. Over 15,000 people died and more than 2500 people are still missing. The displaced population is estimated at around half a million and the damage at around US$200 billion (Oskin 2017) (Fig. 5.3).

The region is subject to a tsunami return period of about 40 years (Esteban et al. 2015). This area was already in socio-economic decline due to a shrinking fishing industry, internal migration to other Japanese cities and demographic change.

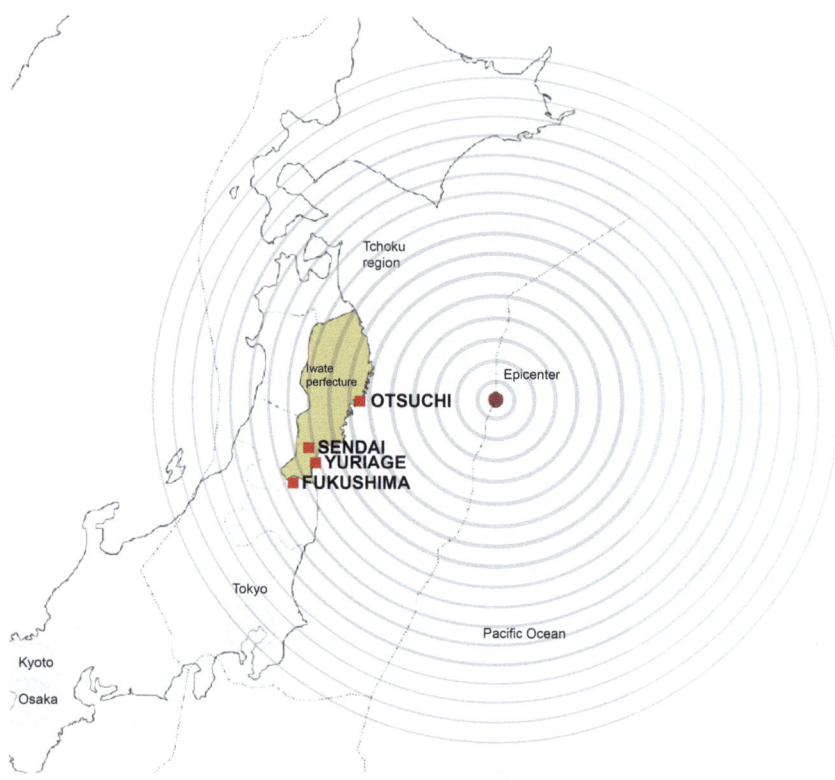

Fig. 5.3   Seismic intensity per region. (Redrawn from Geology Page, 2014)

The tsunami destroyed most coastal villages, including Yuriage and Ötsuchi. Yuriage is a coastal village, part of Natori, on the Sendai plain in Miyagi prefecture. Almost 1000 residents of Natori lost their lives and around 80% of the houses washed away (Murakami et al. 2012) (Fig. 5.4).

Ötsuchi is a coastal village of about 10,000 inhabitants in Iwate prefecture, located between steep mountain slopes. The disaster took the lives of 1281 people (Nakai 2013), while a built-up area of 216 ha was destroyed. Machikata, the central district, was severely damaged. Figure 5.5 shows aerial photos of this district before and after the tsunami.

The government of Japan issued guidelines for the reconstruction shortly after the disaster, emphasizing the need for a comprehensive

Fig. 5.4 Destruction of Yuriage 2008 (left) and 2011 (right) after the great Tohoku tsunami 2011 (Google Earth)

Fig. 5.5 Destruction of Ōtsuchi town by the 2011 tsunami

approach while giving the municipalities a leading role in the reconstruction process (Tanaka et al. 2012).

5.3.1.1 *Yuriage—Reconstruction Measures*
Three categories of tsunami defence measures are constructed in Yuriage (Fig. 5.6):

1. Physical defences: a coastal levee, coastal forests, and elevated roads
2. Relocation: moving residents to raised inland areas
3. Evacuation: vertical evacuation facilities and evacuation routes

The former centre was raised by 4–6 m of sand to create a 'Level 1' flood-safe place to live. New housing blocks were built there and on top

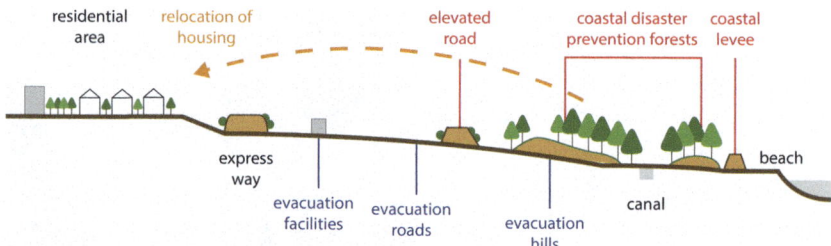

Fig. 5.6 Proposed measures for tsunami defence in Yuriage

of the highest blocks, evacuation shelters were realized. Industrial premises were planned on the lower grounds, between the coastal defence line
and the raised residential area—'Level 2'. Vertical evacuation facilities will
also be created in the industrial zone, as well as elevated roads for horizon-
tal evacuation (TUD 2018).

5.3.1.2 Ötsuchi—Planning Process

During the reconstruction planning of Ötsuchi town, a bottom-up
approach was applied (Nakai 2013). The residents had to take responsibility for initializing the reconstruction process, as the top layer of local government had sadly lost their lives (Takezawa and Barton 2016). Their
planning activities were supported by external consultants (Fukushima
2017). Advisers for Machikata district were academics in the field of spatial
planning. They, however, were also forced to manage civil engineering
design issues, due to shortage of professional civil engineers, while fast
redevelopment was necessary, as many people were relocated.

5.3.1.3 Ötsuchi—Reconstruction Measures

The reconstruction plan is the outcome of an interdisciplinary planning
process between residents and expert advisers. The plan includes the following measures, as shown in Fig. 5.7 (Broere et al. 2019; TUD 2020):

- Construction of a 14 m high seawall and floodgates
- Raising a 31 ha residential area by 2.2 m, for Level 1 protection
  ('Reclamation area')
- Creating a retention area with Level 2 protection for energy
  dissipation

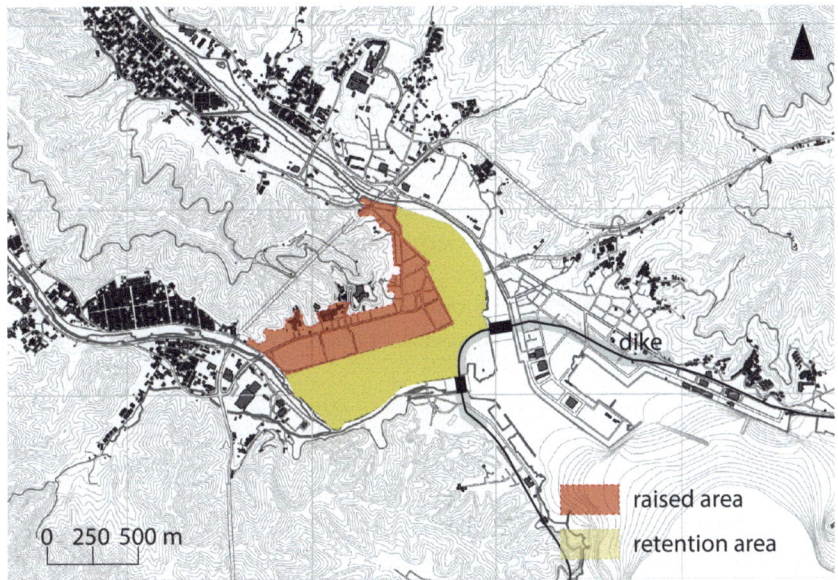

Fig. 5.7 Reclamation plan for Ötsuchi's central district

The result is a multilevel safety approach, as proposed by Iwate prefecture (2011): Retreating from the danger by relocating to higher grounds and raising residential areas, while dissipating the tsunami's energy.

5.3.2 Bahamas

On September 1, 2019, Hurricane Dorian made landfall on the Abaco Islands recording a category 5. On September 2, it hit the island of Grand Bahama with the same force and remained there for another day, finally pulling away from the island on September 3. The wind speeds, up to 295 km/h, also impacted a storm tide of 6.1–7.6 m, covering a large part of Grand Bahama with seawater. Dorian also dropped an estimated 0.91 m of rain over the Bahamas.

The total damage in the Bahamas amounted to US$3.4 billion; there were at least 70 deaths in the country, 10 of which were on Grand Bahama; and another 282 people were missing.

The island of Grand Bahama, with 75,000 residents, suffered from multiple aspects: A king tide, storm surges and waves led to coastal flooding which then caused saltwater intrusion into their drinking water aquifer. At least 60% of Grand Bahama was left submerged once Dorian had passed over the island. Extreme rain led to pluvial flooding, and extreme wind speeds and tornado's damaged buildings, infrastructure, trees and forests. There was an island-wide power outage, and an oil refinery was damaged. About 300 homes were destroyed or severely damaged. Floodwaters and sewage contaminated Rand Memorial Hospital and destroyed supermarkets. The income dip, population decrease and post-traumatic stress were substantial societal problems as secondary effects of the hurricane (Fig. 5.8).

Grand Bahama can be divided into three zones: east, west and the town of Freeport. Freeport is a 560 km² free trade zone on Grand Bahama Island. The Grand Bahama Port Authority (GBPA) operates the free trade zone, under special powers conferred by the government under the Hawksbill Creek Agreement, which was recently extended until August 3, 2054. GBPA and the national authorities are currently developing plans for the recovery of the island. The University of the Bahamas (UB) and TU Delft started assisting in this recovery process with an interdisciplinary design project. Some of the results are summarized in the following subsections.

5.3.2.1 Reconstruction Measures

The main vision for the reconstruction of Grand Bahama is to Build Back Better. This is done by taking an interdisciplinary approach and connecting engineering to spatial planning and design.

The proposed strategy reduces the risk by taking into account exposure and vulnerability of the general risk approach. The main point of the strategy is to create a resilient urban environment in which vital infrastructure like the airport remains operational. This is done by making a collective protection zone of the economic and social city centre of Freeport, a zone that also offers shelter. Individual protection and evacuation shelters will be given to residents, buildings and facilities in the less densely built areas east and west of the city.

The strategy has three intervention layers: hydraulic engineering; zoning and critical infrastructure; and resilient water supply. The hydraulic engineering layer proposes strategic dikes that connect natural heights in the town of Freeport. This creates a zone that is safe from storm surges by

Fig. 5.8 Image of the effects of Dorian on the UB Campus on Grand Bahama Island. (Photo: Fransje Hooimeijer)

a *collective* protection system and an 'outerdike' area that requires *individual* protection systems. Based on this new topography the zoning and critical infrastructure layer proposes interventions per zone. The new safe zone in Freeport (comparable to the traditional mound 'terp' and Dutch polders) will be the centre for critical infrastructure and offers shelter for people from outside this protected zone. Road infrastructure on the island will be improved to give better access to this centre. In the 'outerdike' east and west districts, individual protection needs to be scaled up by building regulations and by creating strategic evacuation centres. The layer of resilient water supply focuses on groundwater resource protection, large-scale rainwater harvesting, the implementation of separate household and drinking water infrastructure, water purification and wastewater treatment facilities (Fig. 5.9).

This vision and strategy is developed in more detailed plans for three locations: the airport, the UB North Outside Campus and the UB town campus. Both the new airport building and the UB North Outside Campus are made out of an ensemble of smaller buildings that are more resistant to the hard winds of the hurricane. They are made flood-proof by lifting them on mounds. The UB In Town Campus is a historical building that is also made flood-proof by placing vital infrastructure like power generators, computers, servers, libraries and toilets on the higher floor. The ground floor is flexibly used. All three buildings are designed as an evacuation point.

Fig. 5.9 The building of strategic dikes (numbers 1 and 3) to create a safe core in Freeport; dikes 4 and 5 are alternatives for number 3

5.4   ANALYSIS

It is interesting to reflect on both cases from the perspective of the principles for Building Back Better. It is important to realize that both cases are incomparable in their phases of recovery. The tsunami hit Tohoku in 2011 and reconstruction activities are well under way, while Hurricane Dorian hit Grand Bahamas in fall 2019 and the recovery process has just started. Hence, in the case of Japan, the study was done 'in hindsight' while in the case of the Bahamas, the situation could be studied in an exploratory way. In Table 5.1, an overview is given of key elements, both in the current situation and in the way the principles can be seen in the results of the study project.

5.5   CONCLUSIONS

As can be seen in all the cases, hardly any efforts are made to match all existing interests in a comprehensive reconstruction plan. Driven by the need for improved safety, the people and the authorities tend to select and implement strong flood defence structures, without added value for everyday use, due to their mono-functional design. 'Build Back Better' is interpreted in a narrow way, focused on flood protection only, and limited to creating threshold protection capacity with levees, seawalls and land raising. Collective protective infrastructural measures on public land are often preferred over individual or hybrid solutions that also involve investment on private property. These collectively protected areas are often exposed to a significant risk of flooding by extreme rainfall; sufficient storm water drainage and storage capacity should therefore be provided. Individual solutions to protect houses, buildings and other infrastructure by a water-robust construction are, if at all, chosen only for areas with very low housing density.

The reconstruction of the destroyed area is hardly, or not, used at all as an opportunity to develop other values, effect changes in the local economy and introduce a more sustainable urban system in terms of drainage, building materials, energy, waste disposal and so on. Preparedness for recovery, reduction of damage sensitivity and strengthening adaptation capacity are not objectives of the reconstruction plan. Vulnerability reduction is most often only achieved by increasing threshold capacity. And whereas raising the ground level is part of the protection strategy in Japan, this is not a typical practice in the Bahamas.

Table 5.1 Guiding principles and interdisciplinarity in the recovery cases in Japan and the Bahamas; current situation and approach in multidisciplinary study project

Physical concepts and interdisciplinary planning approach		Interdisciplinary approach	Japan	Bahamas
Safety first	**Current situation**	Mono-functional separation of safety and spatial planning	Level 1 and Level 2 protection compulsory; realized by a levee, raising the land, strict zoning plan and better infrastructure	Protection strategy is based on evacuation
	Study project	Integration of flood safety and water management in spatial planning	Students followed Level 1 and 2 instructions but balanced the interventions with other spatial programs like natural area	Students proposed to make a safe haven in Freeport, plus amendments to buildings and better infrastructure both inside and outside this protected area
Risk approach	**Current situation**	Risk approach independent of spatial planning; no related spatial planning practice	Hazard reduction is impossible for tsunamis; large emphasis on evacuation and preparedness to save lives and reduce consequences	Hazard reduction is impossible for hurricanes; Hurricane risk is accepted; Dorian is seen as a wakeup call to reduce damage sensitivity
	Study project	Spatial planning based on reduction of consequences and saving lives. Disaster is an opportunity to Build Back Better	The projects basically follow the risk approach of the reconstruction plan but expand it with the connection to spatial and social quality	The study introduces spatial planning integrated with engineered collective flood defense in the center zone, and individual protection and improvement of evacuation in the unprotected areas

(*continued*)

Table 5.1 (continued)

Physical concepts and interdisciplinary planning approach		Interdisciplinary approach	Japan	Bahamas
Capacities for vulnerability reduction	**Current situation**	Spatial planning independent of vulnerability reduction	Reconstruction only focusing on creating threshold capacity	There is no plan; people are coping with the situation
	Study project	Including vulnerability reduction as part of spatial planning is the specific aim of the study	Study introduces Coping, Recovery and, Adaptive capacities by focusing on reduction of consequences and strengthening socio- economic sectors through spatial planning	Study strengthens coping and recovery capacity, Adaptive capacity is improved on the architectural scale by building regulations
3-point approach	**Current situation**	Missing in both cases	Strong focus on standards for threshold capacity / protection levels; mono-functional solutions; no additional measures for adding value or reducing sensitivity	Focus on individual protection without specific standards, and coping capacity in a responsive way
	Study project	Introduce multi-functionality and damage sensitivity in order to achieve three-point approach	Technical optimization is integrated with spatial planning. Every-day added value is especially addressed in Otsuchi	Technical optimization is integrated with spatial planning. Every-day added value is addressed in the protected area center, while sensitivity reduction characterizes the plans for the airport and unprotected area

(continued)

Table 5.1 (continued)

Physical concepts and interdisciplinary planning approach		Interdisciplinary approach	Japan	Bahamas
Four ways to reduce flood risk	**Current situation**	Improved drainage and preparedness are common components for flood protection, levees; land level changes and adapted building are not	Tsunami preparedness is high, Raising land level is essential component of protection strategy, buildings are not adapted to flood risk	Preparedness is the essential protection strategy, local examples of improved drainage, buildings are improved to avoid wind damage, not for flood damage protection
	Study project	Spatial planning and building codes as integrated part of disaster risk reduction	In addition to the reconstruction plan drainage infrastructure improvements and flood resilient building practices are proposed	Improvements are proposed in drainage infrastructure, land level and flood-resilient building
Hybrid solutions	**Current situation**	Combinations of grey, high-tech solutions with nature-based solutions. Focus on reconstruction of housing	Flood walls and levees are in some places combined with coastal forests for impact reduction. Limited attention for reconstruction of economic activities	Forest are applied for impact reduction; grey solutions for flood protection hardly or not applied; reconstruction of economic activities is a private responsibility
	Study project	Exploring combinations of grey, high-tech solutions with nature-based solutions	In addition to the reconstruction plan nature-based solutions are applied for urban drainage and ecological recovery	Nature-based solutions are applied for urban drainage, storm water retention and rainwater harvesting

(*continued*)

Table 5.1 (continued)

Physical concepts and interdisciplinary planning approach		Interdisciplinary approach	Japan	Bahamas
Scoping	**Current situation**	Scoping was not part of the (mono-disciplinary) planning process	There is no practice in Japan that integrates engineering with spatial planning principles or processes, spatial planning is engineering-oriented	There is no practice in the Bahamas that integrates engineering with spatial planning principles or processes, Planning on Grand Bahamas is done by a private party (GPPA)
	Study project		Study was based on scoping	Study was based on scoping
Charette	**Current situation**	Charrette-approach was not part of the planning process	There is no practice in Japan that uses charrettes for multidisciplinary collaborative planning; reconstruction planning is top down, making local stakeholder choose from unintegrated technical solutions	There is no practice in Bahamas that uses charrettes or other forms of co-creative comprehensive planning; Reconstruction is considered to be a private responsibility, with some top-down activities
	Study project		Study was based on charrette to produce a comprehensive plan	Study was based on charrette to produce a comprehensive plan

Comprehensive spatial planning requires the involvement of a multidisciplinary team of experts and local stakeholders. In a collaborative design process, a vision for the future of the recovered community has to be built. Scoping can be used to translate this vision into an integrated reconstruction plan. This process, however, seems difficult to organize due to the existing national spatial planning infrastructure. Moreover, authorities are under pressure to reconstruct housing for the survivors, but it is equally important to reconstruct factories, industries, warehouses and shops, so that survivors can get jobs, an income and access to supplies of food, building materials and so on. Moreover, the survivors need a social shelter point to overcome the pain of the trauma that is engraved in their minds. To Build Back Better, physical reconstruction and the social recovery of communities must go hand in hand.

Acknowledgements This work was funded by the Delta Infrastructure and Mobility Initiative (DIMI) of the Delft University of Technology for the research in Japan and by the University of the Bahamas (UB) for the research in Grand Bahama. The UB dean Carlton Watson was also a great support and partner in the project content-wise. We thank our partners Tohoku University, Tokyo, and Waseda University, International Research Institute of Disaster Science (IRIDS), Municipality of Sendai, ORAGA, and Otsuchi Yume Hiroba for the support in the Japanese cases. Last of all we would like to acknowledge the important contribution to the projects by the staff comprising Jeremy Bricker, Adam Pel, Amin Askarinejad, Robert Lanzafame, Job Schroën and all the students involved.

References

Areso Rossi, A., van Overstraten-Kruijsse, F., Oosterom, M., Moncrieff, N., Suijkens, S., & Grigoris, X. (2018). *Transferring inter-disciplinary flood reconstruction responses from Japan to The Netherlands*. Student report, Delft University of Technology, The Netherlands. Retrieved from: http://resolver.tudelft.nl/uuid:09295919-66cc-42eb-9955-03ffdef1ce83

Blom, M., Das Sharma, A., Houtzager, D., van Klaveren, W., Leung, R., Liu, D., Nguyen, T., Ozcan, A., Prida Guillen, A. (2019). *Conserving Coastal Lagoons by Enhancing Ecosystem Services: A Case Study of the Muni-Pomadze Lagoon in Ghana*. Student report, Delft University of Technology, the Netherlands.

Broere S., Flores Herrera, E., Gori, A., Ozcan, A., Panayi, Z., Prida Guillén, Á. Nimmi Sreekumar, N., & van Unnik, E. (2019). *Interdisciplinary resilient spatial planning based on the reconstruction of Otsuchi, Japan*. Student report, Delft University of Technology, The Netherlands. Retrieved from: http://resolver.tudelft.nl/uuid:5ad3ec97-4fb9-40d0-a9b2-7fbc5eaccc55

De Graaf, R., van de Giesen, N., & van de Ven, F. (2009). Alternative water management options to reduce vulnerability for climate change in the Netherlands. *Natural Hazards, 51*, 407. https://doi.org/10.1007/s11069-007-9184-4.

Dobbelsteen, J. (2018). *The path towards Modern Urban Renewal: Adaptive reconstruction process after tsunami disaster in coastal cities of Japan*. Master's thesis,

Delft University of Technology, The Netherlands. Retrieved from: http://resolver.tudelft.nl/uuid:ba0823d6-20bf-4f62-8214-aab0474b86d8

Duijvestein, C. A. J., & van Dorst, M. J. (2006). Concepts of sustainable development: Brundtland in the built environment – The fourth P. In V. Wang, Q. Sheng, & C. Sezer (Eds.), *Modernization and regionalism. Re-inventing urban identity* (Vol. II, pp. 677–680). Delft: International Forum on Urbanism.

Esteban, M., Takagi, H., & Shibayama, T. (2015). *Handbook of coastal disaster mitigation for engineers and planners*. Elsevier. https://doi.org/10.1016/C2013-0-12806-1.

Filipouskaya, N. (2019). *Experimental investigation of submarine landslide induced Tsunami waves*. Master's thesis, Delft University of Technology, The Netherlands. Retrieved from: http://resolver.tudelft.nl/uuid:d1183805-cb72-48b9-85e7-8459948ce7a8

Fratini, C. F., Geldof, G. D., Kluck, J., & Mikkelsen, P. S. (2012). Three points approach (3PA) for urban flood risk management: A tool to support climate change adaptation through transdisciplinarity and multifunctionality. *Urban Water Journal, 9*(5), 317–331. https://doi.org/10.1080/1573062X.2012.668913.

Fukushima, H. (2017). Workshop arrangement in NeighborHood communities for residents' participation in the reconstruction land readjustment project of Machikata District, Otsuchi Town, Kamihei County, Iwate Prefecture. *Journal of JSCE, 5*(1), 190–205. https://doi.org/10.2208/journalofjsce.5.1_190.

Geology Page. (2014). *Stanford scientists identify mechanism that accelerated the 2011 Japan earthquake* [online]. Available at: http://www.geologypage.com/2014/12/stanford-scientists-identify-mechanism-that-accelerated-the-2011-japan-earthquake.html. Accessed 15 June 2018.

Glasbergen, T. (2018). *Characterization of incoming tsunamis for the design of coastal structures: A numerical study using the SWASH model*. Master's thesis, Delft University of Technology, The Netherlands. Retrieved from: http://resolver.tudelft.nl/uuid:ad229966-5403-432d-9a13-84a9e7fdb5bc

Höller, L., & van de Wiel, T. (2019). *Dismantle boundaries create synergies. Rethinking Houston infrastructure*. Student report, Delft University of Technology, the Netherlands.

Hooimeijer, F., Bricker, J. & I Iluchi (2018) An interdisciplinary approach to urban reconstruction after the 2011 Tsunami. In *TU Delft DeltaLinks*. 12, 1, 11 p.

IPCC. (2007). *Climate Change 2007: Impacts, Adaptation and Vulnerability. Contribution of Working Group II to the Fourth Assessment Report of the Intergovernmental Panel on Climate Change*. M.L. Parry, O.F. Canziani, J.P. Palutikof, P.J. van der Linden and C.E. Hanson, Eds., Cambridge University Press, UK, p 881 and 883.

IPCC. (2014). *Annex II: Glossary* [Mach, K.J., S. Planton and C. von Stechow (eds.)]. In Climate Change 2014: Synthesis Report. Contribution of Working Groups I, II and III to the Fifth Assessment Report of the Intergovernmental Panel on Climate Change [Core Writing Team, R.K. Pachauri and L.A. Meyer (eds.)]. IPCC, Geneva, Switzerland, pp. 117–130.

Iwate Prefecture. (2011). *Iwate Prefecture Great East Japan Earthquake and Tsunami 'reconstruction plan: Basic reconstruction plan* (Tech. Rep.). Iwate Prefecture.

Lennertz, B., & Lutzenhiser, A. (2014). *The charrette handbook. The essential guide for accelerated collaborative community planning* (2nd ed.). Chicago: The American Planning Association. ISBN 9781611901474.

Li, Y., Dil, K., & van Unnik, E. (2019). *Humanizing Houston. construction & water resilient design of downtown Houston.* Student report, Delft University of Technology, The Netherlands.

Möhring, R. (2018). *Sustainable design of transport systems: A transport design strategy in response to the Great East Japan Earthquake considering the trends of Shrinking Cities and the Aging Society.* Master's thesis, Delft University of Technology, The Netherlands. Retrieved from: http://resolver.tudelft.nl/uuid:c8914b31-8ef5-45a5-a41b-e6af4ae42bba

Mujumdar, G. (2019). *KiNTSUGi: Improving resilience capacities in a hazardscape, Otsuchi, Japan.* Master's thesis, Delft University of Technology, The Netherlands. Retrieved from: http://resolver.tudelft.nl/uuid:ff2bfd11-9edb-42d3-a283-781871ba9640

Murakami, H., Takimoto, K., & Pomonis, A. (2012). *Tsunami evacuation process and human loss distribution in the 2011 Great East Japan Earthquake: A case study of Natori City, Miyagi Prefecture.* Available at: https://www.iitk.ac.in/nicee/wcee/article/WCEE2012-1587.pdf. Accessed 13 June 2018.

Mustaqim, M. (2018). *Stability analysis of geotextile-reinforced slope based on Japan earthquake in 2011: Yuriage, Natori City case.* Master's thesis, Delft University of Technology, The Netherlands. Retrieved from: http://resolver.tudelft.nl/uuid:5f83f358-2ef2-4d31-971e-cc8a3011d39d

Nakai, Y. (2013). Reconstruction plan of Otsuchi Town, Kamihei County, Iwate Prefecture. *Journal of JSCE, 1*(1), 242–250. https://doi.org/10.2208/journalofjsce.1.1_242.

Nederlof, I. (2019). *Towards Resilient Urban Stormwater Management in a Tsunami Reconstruction: A Scenario Discovery Study on Ötsuchi Town, Japan.* Master's thesis, Delft University of Technology, The Netherlands. Retrieved from: http://resolver.tudelft.nl/uuid:191f7a6e-e7d7-4725-b319-579da93ab265

Oskin, B. (2017). *Japan Earthquake & Tsunami of 2011: Fact and information* [Online]. Available at: https://www.livescience.com/39110-japan-2011-earthquake-tsunami-facts.html. Accessed 7 May 2018.

Prida Guillén, Á. (2019). *Evaluation of the feasibility of solutions to flash flooding in the municipality of Tirana (Albania)*. Student report, Delft University of Technology, the Netherlands.

Rao, A. (2019). *Stitches: Blending landscape fabric through the golden threads of spatial identity in San Riku coastline, Otsuchi, Iwate, Japan*. Master's thesis, Delft University of Technology, The Netherlands. Retrieved from: http://resolver.tudelft.nl/uuid:e592e197-a92e-47fa-8f54-1a17b239a204

Roubos, J. (2019). *Prediction of the characteristics of a tsunami wave near the Tohoku coastline: Numerical SWASH modelling*. Master's thesis, Delft University of Technology, The Netherlands. Retrieved from: http://resolver.tudelft.nl/uuid:421cd6b8-fd31-424a-aa9b-529dc17018eb

Salet, J. (2019). *Tsunami induced failure of bridges: Determining failure modes with the use of SPH-modeling*. Master's thesis, Delft University of Technology, The Netherlands. Retrieved from: http://resolver.tudelft.nl/uuid:56d1ccc2-b50a-44a6-9908-d5f495c6951c

Sugano, K., & Lu, S. (2019). *Hybridity vs Closed City: A study about the impact of applying "Hybridity" as a concept of understanding in designing a decentralized water circulation urban model called "Closed City"*. Report, Delft University of Technology. http://resolver.tudelft.nl/uuid:08eb35f4-634f-4ef9-a3fb-fe48f700849f

Takezawa, S., & Barton, P. T. (2016). *The aftermath of the 2011 East Japan earthquake and tsunami: Living among the rubble*. Lanham: Lexington Books.

Tanaka, Y., Shiozaki, Y., Hokugo, A., & Bettencourt, S. (2012). *Reconstruction policy and planning*. Washington, DC: World Bank.

TUD. (2018). Reconstruction of Yuriage [video file]. Retrieved from: https://d1rkab7tlqy5f1.cloudfront.net/Websections/Infrastructures%20and%20Mobility/Resilient%20and%20Adaptive%20Urban%20deltas/DIMI%20V6%20Short.mp4

TUD. (2020). OTSUCHI emergency planning for resilience [video file]. https://youtu.be/OoANpXJsxT4

UNISDR. (2005). *Hyogo framework for action 2005–2015*. Retrieved from: https://www.unisdr.org/2005/wcdr/intergover/official-doc/L-docs/Hyogo-framework-for-action-english.pdf

UNISDR. (2015). *Sendai framework for disaster risk reduction 2015–2030*. Retrieved from: https://www.undrr.org/publication/sendai-framework-disaster-risk-reduction-2015-2030

UNISDR. (2017). *Build back better – In recovery, rehabilitation and reconstruction* (Tech. Rep.). United Nations Office for Disaster Risk Reduction. Retrieved from https://www.unisdr.org/files/53213_v.bbb.pdf

Vafa, N. (2018). *Activate resilience of the Miyagi coast*. Master's thesis, Delft University of Technology, The Netherlands. Retrieved from: http://resolver.tudelft.nl/uuid:9200c56a-e0c8-4e9e-b5b3-38e69511c49f

Van den Berg, N., Hendriks, O. & Boertje, L. (2019). *Midtown: Right in the MIDst of a climate resilient and a vibrant housTOWN.* Student report, Delft University of Technology, the Netherlands.

Van de Ven, F., Luyendijk, E., & de Gunst, M. et al. (2009). Waterrobuust Bouwen. (Water-robust urban development; in Dutch), Beter Bouw- en Woonrijp Maken. ISBN 978 90 5367 496 3.

van Dijk, M. (2018). *Tsunami resiliency of transport systems: The development and application of a tsunami resiliency assessment method.* Master's thesis, Delft University of Technology, The Netherlands. Retrieved from: http://resolver.tudelft.nl/uuid:e0ea6263-c9fd-444d-8176-56acef2c9601

Yasaku, Y. (2019). *Extensive application of a methodology to evaluate a tsunami-resilient transportation system.* Master's thesis, Delft University of Technology, The Netherlands. Retrieved from: http://resolver.tudelft.nl/uuid:bbe040ae-44f2-44d2-a4e4-867b6c7c2d4c

CHAPTER 6

Removing Challenges for Building Resilience with Support of the Circular Economy

Jeroen Rijke, Liliane Geerling, Nguyen Hong Quan, and Nguyen Hieu Trung

Abstract This chapter explores the concepts of resilience and circular economy to see if they can complement each other. Whilst resilience thinking is focused on vulnerability, system adaptation and transformation, circular economy is oriented at optimizing value from the (natural) resources

---

For translations and insights into Vietnamese culture we got help from Ms. Nguyen Thi Thanh Hue from WACC and Võ Nguyễn Minh Thùy, student at CTU, as well as from some other students from CTU. Furthermore, we were assisted by students from VHL UAS, HZ UAS and HAN who collected data during their internship or final thesis project: Luuk van den Ham, Martijn Genet, Jelle Janssen, Kornelis Kramer, Cor Lange, Luuk van der Meulen, Sjoerd van der Meulen, Roy Ruiter and Heleen Sijses.

J. Rijke (✉)
Sustainable River Management Group, HAN University of Applied Sciences, Arnhem, The Netherlands

VHL University of Applied Sciences, Velp, The Netherlands
e-mail: j.rijke@han.nl

that are available to societies. We use the city of Can Tho in the Vietnamese Mekong Delta as a case study and demonstrate that a circular plastic approach can be supportive for implementing the resilience strategy. It is found that the use of floating structures for agriculture and aquaculture made of recycled (river) plastic could benefit the livelihoods of local people in terms of minimizing damage and casualties caused by river erosion and seasonal flooding on the one hand, and adding extra income and food security on the other. Including the concept of circular economy may contribute to the adaptive capacity of flood-prone communities in the Vietnamese Mekong delta and as such enhance their resilience.

Keywords Circular economy • Resilience challenges • Can Tho • Vietnamese Mekong Delta

6.1   Introduction

The concept of resilience supports planners and policymakers by addressing inherent uncertainties of complex social–ecological systems (Folke 2016; Davoudi 2012). It has gained a central position in many adaptation policies and plans, although an implementation gap remains, as progress of adaptation falls behind societal needs and ambitions, due to a variety of reasons that relate to multiple factors, such as finance, governance, technology, societal behaviour and politics (UN Water 2020; GCA 2019).

Resilience thinking, which has a focus on vulnerability, system adaptation and transformation, also provides a lens on sustainability (Folke

L. Geerling
Department of Technology, Water and Environment, HZ University of Applied Sciences, Middelburg, The Netherlands

N. H. Quan
Institute for Circular Economy Development (ICED), Vietnam National University – Ho Chi Minh City, Ho Chi Minh City, Vietnam

Center of Water Management and Climate Change (WACC), Institute for Environment and Resources (IER), Vietnam National University – Ho Chi Minh City, Ho Chi Minh City, Vietnam

N. H. Trung
Climate Change Research Institute (DRAGON-Mekong), Can Tho University, Can Tho, Vietnam

2016). Although the emerging literature on the development of resilience pays explicit attention to poverty traps and adaptive capacity (e.g. Barrett and Constas 2014; Béné et al. 2016), resilience narratives have been accused of reframing policy issues in a way that makes vulnerable populations responsible for securing themselves (Tanner et al. 2017). Such anomalies may be avoided by mirroring resilience thinking with other sustainability perspectives.

This chapter explores whether the concept of circular economy (CE), which is grounded in ecological economics and industrial ecology, can be used to overcome challenges for establishing urban resilience in practice. Geissdoerfer et al. (2017) defined CE as "a regenerative system in which resource input and waste, emission, and energy leakage are minimised by slowing, closing, and narrowing material and energy loops. This can be achieved through long-lasting design, maintenance, repair, reuse, remanufacturing, refurbishing, and recycling. Second, sustainability is defined as the balanced integration of economic performance, social inclusiveness, and environmental resilience, to the benefit of current and future generations." CE is almost exclusively developed by business practitioners (Korhonen et al. 2018) and adopts design and business perspectives to sustainability for designing out waste and pollution, keeping materials in use and regenerating natural systems (Ellen MacArthur Foundation 2017). It is worthwhile exploring whether CE can be complementary to resilience and whether they can be combined to overcome implementation challenges for climate adaptation. For this, the city of Can Tho in the Vietnamese Mekong Delta (VMD) is used as a case study.

6.2   Can Tho Resilience Strategy, Potential for Circular Economy?

With about 1.3 million inhabitants, the city of Can Tho is the largest city in the VMD. The VMD, which extends over the vast area of southern Vietnam, is considered one of the world's most sensitive areas to climate change (WWF 2009). The city of Can Tho is situated at the interplay of the Can Tho River, the Binh Thuy River and the Hau River, approximately 80 km from the Hau River mouth (Fig. 6.1). The Hau River is one of the main branches of the Mekong River, whose water levels are significantly influenced by the ocean tides in addition to seasonal high river discharge and consequent inundations (Takagi et al. 2014). People in the

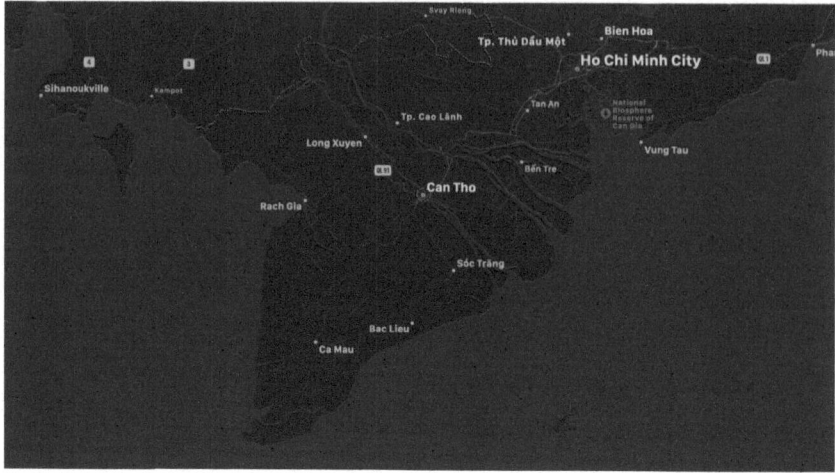

Fig. 6.1   Location of Can Tho in the VMD. (Source: own drawing Geerling 2020)

VMD have been subjected to the impacts of flooding and inundation. Living conditions and livelihoods in the VMD were historically well adapted to the regular pattern of seasonal flooding, which residents and local governments describe as a "living-with-floods" strategy. However, during the past few years, flooding has become less predictable and more damaging due to a multitude of factors such as climate change, sea level rise, subsidence (due to groundwater extraction) and urban development (People's Committee of Can Tho City 2019).

Can Tho participated in the Rockefeller Foundation's 100 Resilient Cities Centennial Challenge. Between May 2016 and July 2019, the city exchanged knowledge and experiences with other cities that participated and developed their Can Tho Resilience Strategy (CTRS). The Cities Resilience Framework (CRF) (Arup 2015) provided guidance for developing the strategy. The CRF defines urban resilience as "the capacity of cities to function, so that the people living and working in cities – particularly the poor and vulnerable – survive and thrive no matter what stresses or shocks they encounter" (ibid).

As a first step towards the CTRS, a vulnerability assessment was made. As a result, it was identified that the most important present-day shocks are flooding, extreme rainfall, extreme heat waves, saline intrusion, environmental incidents and high tides. And land subsidence, reduction of green and blue space, environmental degradation, ageing infrastructure, legal and technical barriers of export markets, migration, poverty and unemployment were identified as the most important ongoing stresses.

The CTRS sets four goals until 2030, corresponding to the four dimensions of the CRF (People's Committee of Can Tho City 2019, p. 27):

- *Leadership and Strategy:* Policies and plans are developed and implemented in a systematic, integrated manner, with active participation of all relevant stakeholders.
- *Infrastructure and Environment:* A green and sustainable river city with an infrastructure system that is well connected, modern, flexible, diverse and resilient to extreme natural hazards.
- *Economy and Society:* A knowledge economy that is proactive, diverse and deeply integrated, while remaining steady when facing regional and global economy fluctuations.
- *Health and Well-being:* Communities have a secured and stable income, and live in a green and clean environment, safe from the impacts of economic, social and environmental shocks and stresses.

The CTRS lists a range of actions to achieve each of these goals. In addition, for each dimension it also highlights the most pressing limitations and challenges. These are summarized in Table 6.1.

Although the concept of CE is not addressed in the CTRS, the waste challenge within the city is mentioned: "The amount of domestic solid waste generated is about 930 tonnes/day (in 2016) and is expected to reach 2,000 tonnes/day in 2030. However, the city's waste and wastewater treatment capacity is still very limited despite considerable efforts."

River deltas act as conveyor belts from land to ocean of natural materials such as water and sediment, but also of (plastic) litter (Jambeck et al. 2015). When observing public space and the river and canal system in the VMD in general and in Can Tho specifically, it becomes apparent that large amounts of litter are not discarded in an environmentally safe manner and end up in public spaces and (directly or indirectly) in the water system. This leads to environmental degradation, which might increase in the future due to the predicted population growth, urban development, industrial development and water scarcity. A big part of the floating litter is composed of (single-use) plastics. Currently, Vietnam ranks 4th (ibid.)

Table 6.1 Limitations and challenges for building resilience in the city of Can Tho (People's Committee of Can Tho City 2019)

Resilience dimension	Limitations and challenges
Leadership and strategy	Lack of integrated cross-sectoral planning process
	Ineffectiveness in the implementation of policies and plans
	Limited capacity to support integrated planning and policymaking
	Lack of stakeholder participation
Infrastructure and environment	Low capacity to respond to extreme events
	Inaccessible and outdated database
	Decreasing green and blue space coverage
	Increasing impacts of flooding
	Poorly connected transport system, and poor-quality public transport system
	Improper solid waste and wastewater management
	Lack of water security
	Insecure energy
Economy and society	Improper planning process
	Lack of linkages between actors in value chains
	Diversification of key agriculture products
	Tariff and technical barriers of export markets
	Improper post-harvest processing
	Lack of regional connectivity
Health and well-being	Ineffective quality of vocational training, livelihood improvement and post-training employment support
	Improper livelihood support to communities following a shock
	Lack of responsive/backup plans for electricity, water and health care service disruptions

and the Mekong River 11th in the top 20 worldwide rivers' contribution to the plastic soup into the world's oceans (Lebreton et al. 2017).

The impacts of plastic pollution are wide-ranging, affecting the riverine environment and canals in Can Tho locally but eventually the marine environment downstream, the East Sea and beyond. Businesses in Can Tho dependent upon clean water and a (plastic) litter-free environment such as fisheries, tourism and food and beverage companies may experience economic losses. Disaster risk may rise due to increased flooding from plastic litter congesting waterways. Although not yet comprehensively researched, (toxic) plastics that enter the water and food chain might impact human health, as well as cause an increase in waterborne diseases (Mathews and Stretz 2019).

6.3 METHODOLOGY

From November 2018 to November 2019 a team of Vietnamese and Dutch researchers explored whether the concept of a circular plastics economy could contribute to addressing the limitations and challenges mentioned in Table 6.1. We focused on macro-plastic litter (> 5 cm) in the river and canal system in and around Can Tho and based our consecutive steps on (a simplified version of) the source-to-sea approach, which consists of the six steps that are described in Table 6.2 (Mathews and Stretz 2019).

Subsequently, how a circular plastics economy could strengthen urban resilience in Can Tho was analysed. This was done by reflecting upon the five capacities of resilience as introduced in Chap. 1 and the resilience-building challenges and limitations that were described in Table 6.1.

6.4 RESULTS

6.4.1 Characterize: Amount and Type of Floating Debris

Via Can Tho's Department of Construction, five companies are responsible for the formal waste collection. They collect approximately 930 tonnes of solid waste per day (People's Committee of Can Tho City 2019). A study from 2011 indicates that this amount consists of approximately 10% of plastic and rubber (Nguyen and Le Hoang 2011). An unknown amount of recyclable plastic is collected by the informal sector.

The water system and riverbanks in and around Can Tho are polluted, as shown in Fig. 6.2. Measurements with a provisional passive litter trap of 4 m² along Can Tho River on different days and times gave a first indication of the composition of the floating debris (see Table 6.3). Of the trapped floating material, 55% consisted of water hyacinth. Of the non-organic material, 85% consisted of plastics, mostly low-density polyethylene (LDPE, (food)-packaging material and wrapping foil), polyethylene terephthalate (PET) and (expanded) polystyrene ((E)PS). Systematic

Table 6.2   Research activities structured by the source-to-sea approach

Step	Explanation
1. Characterize	With desk research, visual observations of the rivers and canals in and around Can Tho (upstream and downstream) and installing a temporary passive litter trap in Can Tho River, a first insight into the types and amount of plastic litter was obtained. Also the flow velocity and movement of plastic litter through the water was measured via a tracker in a plastic bottle. To have a better insight into locations along the Hau River that suffer from erosion and flood risk a desk research was executed and information was gathered via the water resources and the civil engineering departments of CTU
2. Engage	Data for this step were collected through a series of face-to-face interviews with individuals who held formal and informal positions across the waste management sector (from collection to sorting and recycling) in Can Tho and around: 10 informal waste collectors, 5 informal waste collection points on neighbourhood level, 2 shredding facilities on city level and 1 formal Public Urban Environment Company that collects household waste for the city government. Next to that, 60 semi-structured interviews were conducted among fishermen owning floating fish cages on different locations along the Hau River and among inhabitants impacted by river erosion at different sites between Long Xuyen and Can Tho. Questions concerned effects of climate change, erosion, flooding, livelihood and potential interest in floating agriculture and/or aquaculture. Questioning and answering occurred in Vietnamese, translations were made into English via an interpreter and then notes were taken and verified
3. Diagnose	The potential value chain for (floating) plastic litter was described based on steps 1 and 2
4. Design	To discuss which approach could contribute to prevent plastic leakage, and which social, environmental and/or economic benefits could be reaped by the different stakeholders in the VMD, two local seminars were organized (November 2018) and a presentation to the People's Committee of Can Tho (November 2019). Diverse interventions to end plastic leakage in and around Can Tho were demonstrated and discussed. Questioning and answering occurred in Vietnamese, translations were made into English via an interpreter and then notes were taken and verified
5. Act	As an intervention to prevent plastic leakage in Can Tho a location study for the implementation of two passive litter traps in Can Tho was executed. In addition, products made of recycled (river) plastic that contribute to the prevention of river erosion and/or adaptation to flooding were discussed with potential users
6. Adapt	Monitor the outcomes of the interventions and adapt as needed

Fig. 6.2 Polluted riverside along Can Tho River. (Source: own picture Geerling 2019)

Table 6.3 Aggregate catcher yield from 11 samples during different tides, total sampling duration 11 hours (Van den Ham et al. 2019)

Waste type	Dry weight (kg)	Percentage
Water hyacinth	6.41	55.0
LDPE	1.9	16.3
PET/PETE	0.9	7.7
(E)PS	0.75	6.4
HDPE	0.6	5.1
PVC	0.15	1.3
PP	0.1	0.9
Other (rubber, glass)	0.85	7.3
Total	**11.66**	**100**

visual observations suggested similar findings, and also pointed out hotspots of floating debris near markets and in inner-city canals and storm water drains. This means that a big part of the floating plastics consists of low-/no-value single-use plastics. It can be concluded that any project for trapping plastic litter from the water system in the VMD should be able to deal with large amounts of water hyacinth, which in itself forms a perfect, natural passive litter trap.

6.4.2   Engage

Households that were interviewed stated they pay around € 0.80/ month/household for waste collection. This is less than the average in Vietnam which is € 1/month/household and does not cover the costs of waste handling (Van Den Berg and Duong 2018). The formally collected household waste is transported to a landfill or one of the city's four incinerators (waste for energy) with a combined capacity of 800 tonnes per day. Only one of the incinerators complies with European environmental standards. Officially, households are obliged to separate their waste into burnable and non-burnable categories before disposing it into waste containers or on the street, but not many people do this. Households can be fined for incorrect separation of waste, but interviewees indicated that they do not separate waste because of convenience and lack of enforcement.

However, recycling of valuable materials takes place. Employees from the formal waste companies separate recyclables such as plastics, glass, metals and cardboard from the household waste during collection, and bring it to informal recycling shops, which can be found at the neighbourhood scale throughout the city, to gain a bit of extra income. Unfortunately, one of the formal waste collection companies wants to forbid this, as it consumes extra time and reduces the total waste volume they are obliged to collect.

Likewise, informal waste collectors go from door to door to collect recyclables—sometimes even pay a small fee for it—and sell it at the informal recycling shops (Fig. 6.3). Plastics that are accepted are polypropylene (PP), PET, high-density polyethylene (HDPE) and LDPE and prices are around € 0.30, € 0.30, € 0.19, and € 0.04 respectively per kilogramme. These prices only vary slightly per recycling shop and depend on regional market prices, making the revenue for informal waste collectors uncertain. The daily yields of plastics vary between approximately 300 kg and 5000 kg, depending on the size of the waste collection shop. Plastic litter is hardly harvested from the rivers or canals by waste collectors, as many informal recycling shops do not accept it when it looks polluted, or reduce the price by 50%. One formal waste management company indicated that they only collect floating litter in the touristic Ninh Kiều area with a canoe during important events like the lantern festival.

In the informal waste collection shops, a first sorting of plastics takes place on product and plastic type (e.g. buckets, bottles). Next, the collected plastic material is transported to a shredding facility within the city

Fig. 6.3 (a, b) An informal waste collection point at neighbourhood scale. (Source: own pictures Geerling 2019)

(Fig. 6.4). One shredding facility indicated that it receives around 1000 kg of plastic per day. Here, plastics are further sorted by plastic type and colour, shredded, washed and dried before being transported to recycling factories in Ho Chi Minh City, due to the absence of recycling factories in Can Tho. During field visits, no quality control at the informal collection points or shredding facilities was mentioned or observed, other than a visual check during intake from the waste collectors. Owners of the shredding facilities mentioned that they would prefer if they could transport the shredded plastics to a recycling factory nearby to save transport costs of around € 40/truck.

Fig. 6.4 (a, b) A shredding facility in Can Tho. (Source: own pictures Geerling 2019)

The CTRS describes that flooding and riverine erosion due to sand mining, upstream dams and navigation are big risks in the VMD, resulting, for example, in damage to infrastructure and buildings and loss of agricultural land. The majority of interviewed fishermen and people living close to erosion sites along the river stated that erosion is a direct threat to their livelihood. A majority of the fishermen also indicated that they would love to have a piece of land to grow food to support their livelihood. About half of these indicated that a floating island could be a good solution and recycled (river) plastic was predominantly suggested as the best construction material for such an island (Kramer 2019).

6.4.3 Diagnose

In 2018 the Vietnamese government banned the import of material for recycling in an attempt to end global waste streams into Vietnam. This resulted in a rising demand for recyclable plastic waste within the country. Due to the inadequate or absent segregation of waste at source, the informal sector fills this gap but only collects plastic waste that can currently be sold to recycling shops.

Although Can Tho River and Hau River are waterways with heavy navigation, the width of the rivers allows for safe instalment of equipment to catch plastic litter. The tidal effect is significant due to steep river banks, as the water rises and drops about 1.5–2 m every 6 hours. The flow also changes due to the semi-diurnal tides; therefore it would be preferable for passive litter traps to catch litter from two directions. Catching plastics from the river and canal system could also be executed with local boats and a mechanical gripper, as water hyacinth is omnipresent and traps already many plastics. This would eliminate high investment costs. The water hyacinth could among others be used for composting, woodcraft or generating electricity. The plastics that are left over after digestion could subsequently be recycled.

In 2019, the Ocean Cleanup launched a plan to install a self-automated trash-collecting vessel in the Can Tho River. Clear Rivers, a Rotterdam-based non-profit organization, also developed a plan to install two passive litter traps made of recycled (river)plastic in the water system in Can Tho. Both initiatives got permission from Can Tho's People's Committee to prepare for implementation, but were not implemented yet at the timeof writing this chapter. Both initiatives are costly and fully depend on donor organizations for funding. One of the formal waste companies agreed to operate the passive litter traps from Clear Rivers once donated to Can Tho city.

6.4.4 Design

During seminars with local government officials and academics, it was suggested that using recycled plastic as a resource for construction material may be worthwhile, as there is a demand for low-cost building materials. This provides a design challenge to develop safe and durable products against a minimum price.

Another option that was discussed was the application of floating structures, made of recycled plastics, for example floating islands for agriculture and/or aquaculture. For example, Clear Rivers designed a modular floating platform that reuses (E)PS in the core, and can be used for a floating park (applied in Rotterdam), but also for floating agriculture and/or aquaculture. This is a product that fits in the floating culture of the VMD, and was seen as a potential successful product by several local specialists and academics, as well as local fishermen. To design products that could reuse (E)PS would add an extra material that could be collected by the informal collection system. At the moment, high storage and transportation costs often prohibit the collection and recycling of (E)PS waste. A mobile recycling unit placed on a truck which compacts the (E)PS on location and can be used on a regional scale reduces the transportation cost of loose (E)PS materials, but also saves storage space. It is currently being implemented in Malaysia and could also be a successful option for the VMD.

6.4.5   Act

If the floating plastic litter from the river and canal system is retrieved, either passively or actively, before it floats further downstream and added to the informal land-based plastic collection chain in Can Tho, it could secure sufficient feedstock to set up a local recycling plant to produce local products that could contribute to flood-proofing the VMD, among others, for the production of floating platforms for food production, but also for drainage boxes or embankments.

6.4.6   Adapt

The introduction of a circular plastics economy using floating concepts can give multiple benefits. The benefits can include improved water quality by removing plastic (toxic) litter from the river and canal system; motivate local communities to retrieve plastic litter; and create new means for local livelihoods, for example, with floating agricultural and/or aquaculture practices on floating islands made of recycled plastic. This CE-based floating concept provides an excellent example of adapting the current waste flow management in Can Tho city.

To monitor if the amount of waste released into the environment and into waterways will be reduced by planned interventions a target must be set by the city government for the reduction of plastic leakage per year. Besides that, it is also important to include economic benefits in the monitoring. As Can Tho University is already a member of the responsible team for the CTRS, they could help in formulating a reliable and accepted monitoring and reporting structure as well as execute it according to a set time frame. It should be noted that the monitoring of plastic leakage is a challenging task with no standardized approaches available yet.

6.5 Discussion

The results suggest that there are opportunities for enhancing a circular plastics economy in the city of Can Tho. One route is to blend river plastic with material streams in the (in)formal land-based waste management system. Although the national waste management policies in Vietnam do not explicitly mention CE, a waste management vision has been provided by the National Strategy of Integrated Solid Waste Management up to 2025 (PM 2009). This could be used as a regulatory framework for the implementation of CE, because it contains specific targets: "A system of integrated solid waste management shall be built. As a result, solid waste is sorted at sources, collected, reused, recycled and treated completely by suitable advanced technologies, minimizing wastes to be landfilled in order to economize land resources and prevent the environment from pollution." It should be noted that the formal waste management in Can Tho is currently oriented at incineration, meaning that the outcomes in terms of CE would be limited to energy production. However, the institutional and regulatory gap for river plastic potentially provides opportunities for large-scale private initiatives to develop a holistic source-to-sea approach. For example, Ocean Cleanup and Clear Rivers have made first steps towards such an approach in Can Tho.

Another route is to strengthen mechanisms that are already ongoing in the informal sector. For example, connection of plastic waste catchers to the informal system could generate new sources of income for informal waste collectors and generate more material to be recycled. Such a strategy is most likely to become actionable with small-scale catchers and manual catching (a local boat with a gripper), as these require relatively low upfront investments. Depending on the level of pollution and product requirements, this material could be mixed with plastic that is collected from residential areas. Research in Nairobi, Kenya, has demonstrated that such an approach could generate social benefits for waste pickers including increased income level and stability, sense of belonging and respect (Gall et al. 2020). Moreover, the same study revealed that this approach could lead to materials that are comparable to state-of-the-art recycled material obtained from advanced formal recycling systems in high-income countries in terms of both composition and basic engineering properties if the right incentives, trust and state-of-the art technology are present (ibid.).

Hence, particularly the route towards CE via the informal sector provides opportunities for mutually strengthening urban resilience. It will support new livelihood models that are beneficial to water quality, and provide alternative resources for building materials that can be used for enhancing infrastructure and buildings, including amphibious/floating structures. Upon reflection of the CRF (Arup 2015), it becomes clear that the CE could contribute to all six resilience qualities: river cleanup (*resourcefulness*), job creation and additional incomes for vulnerable urban community members (*inclusiveness*), recycled plastic as an alternative for virgin plastic (*resourcefulness*) and application of recycled plastic for floating structures for agriculture and housing (*redundancy, robustness, flexibility*).

Both informal and formal routes provide incentives for stakeholder collaboration and strengthening of linkages across the value chain. In addition, they provide monetary returns for waste management companies (formal route) and/or waste collectors (informal route). This may support the improvement of solid waste management in the city of Can Tho and enhance the effectiveness of implementation of the Can Tho Resilience Strategy (particularly towards the goals of *Infrastructure and Environment* and *Health and Well-being*). With this, CE provides an opportunity to address some of the limitations and challenges for resilience building as outlined in Table 6.1. However, it should be noted that CE is not the solution to all limitations and challenges. It should therefore not be considered as the only approach to tackle resilience challenges and limitations.

The case of Can Tho shows that CE connects to three out of the five pillars for climate resilience that are described in Chap. 1. Whilst CE may not reduce the climate-induced vulnerability of Can Tho city, it enhances its *recovery capacity* by supporting livelihoods of the urban poor. And, consequently, it contributes to tackling a number of challenges for building resilience and thus increases its *transformative capacity* towards implementation of its resilience strategy. Furthermore, it could possibly provide construction material for measures, such as floating structures, that increase the city's *adaptive capacity*.

6.6 Conclusion

Adapting cities to the present and future threats of climate change and enhancing their resilience imply substantial challenges as well as opportunities, determined by the particularities of each urban system. The case study in Can Tho city shows that the concepts of resilience and CE can complement each other and suggests that CE could be used to address a number of resilience-building challenges. In this study, we successfully applied a simplified source-to-sea approach which consists of six steps in Can Tho city. In the case of Can Tho it was found that resilience provides a scientific background, whilst CE contributes to practical implementation. Resilience and CE both have a regenerative nature, but with different perspectives. Resilience focuses on vulnerability, and adaptive and transformative capacity, whereas CE primarily looks at business and design practices. As such, CE could create opportunities for bringing short-term benefits into climate adaptation policies and action.

Cities across the globe are putting deliberate efforts into both becoming resilient and developing CE. However, these efforts are often performed separately. Based on the case of Can Tho, we suggest combining these perspectives for gearing action towards sustainable livelihoods as well as an improved adaptive capacity.

References

Arup. (2015). *City resilience framework.* https://www.rockefellerfoundation.org/report/city-resilience-framework/. Accessed 31 Mar 2020.

Barrett, C. B., & Constas, M. A. (2014). Toward a theory of resilience for international development applications. *Proceedings of the National Academy of Sciences, 111*(40), 14625–14630.

Béné, C., Headey, D., Haddad, L., & von Grebmer, K. (2016). Is resilience a useful concept in the context of food security and nutrition programmes? Some conceptual and practical considerations. *Food Security, 8*(1), 123–138.

Davoudi, S. (2012). Resilience: A bridging concept or a dead end? *Planning Theory & Practice, 13*(2), 299–307.

Ellen MacArthur Foundation. (2017). The new plastics economy: Rethinking the future of plastics & catalysing action.

Folke, C. (2016). Resilience (Republished). *Ecology and Society, 21*(4), 44.

Gall, M., Wiener, M., de Oliveira, C. C., Lang, R. W., & Hansen, E. G. (2020). Building a circular plastics economy with informal waste pickers: Recyclate quality, business model, and societal impacts. *Resources, Conservation and Recycling, 156,* 104685.

GCA. (2019). *Adapt now: A global call for leadership on climate resilience.* Global Commission on Adaptation. https://gca.org/global-commission-on-adaptation/report. Accessed 31 Mar 2020.

Geissdoerfer, M., Savaget, P., Bocken, N. M. P., & Hultink, E. J. (2017). The circular economy – A new sustainability paradigm? *Journal of Cleaner Production, 143,* 757–768.

Jambeck, J. R., Geyer, R., Wilcox, C., Siegler, T. R., Perryman, M., Andrady, A., Narayan, R., & Law, K. L. (2015). Plastic waste inputs from land into the ocean. *Science, 347*(6223), 768–771.

Korhonen, J., Honkasalo, A., & Seppälä, J. (2018). Circular economy: The concept and its limitations. *Ecological Economics, 143,* 37–46.

Kramer, T. K. J. (2019). *Creating sustainable livelihoods in the Mekong delta.* https://hbo-kennisbank.nl/details/sharekit_hz:oai:surfsharekit. nl:66d78f30-91d3-44a4-a56a-89766588757d?q=CREATING+SUSTAINAB LE+LIVELIHOODS+IN+THE+MEKONG+DELTA&has-link=yes&c=0. Accessed 31 Mar 2020.

Lebreton, L. C. M., Van der Zwet, J., Damsteeg, J. W., Slat, B., Andrady, A., & Reisser, J. (2017). River plastic emissions to the world's oceans. *Nature Communications, 8,* 15611.

Mathews, R. E., & Stretz, J. (2019). *Source-to-sea framework for marine litter prevention: Preventing plastic leakage in river basins.* Stockholm: SIWI.

Nguyen, H., & Le Hoang, V. (2011). Solid waste management in Mekong Delta. *Journal of Vietnamese Environment, 1*(1), 27–33.

People's Committee of Can Tho City. (2019). *Can Tho resilience strategy until 2030.* Can Tho. https://www.i-s-e-t.org/cantho-resilience-strategy-100rc. Accessed 24 Apr 2020.

PM. (2009). *Quyết định 2149/2009/QĐ-TTg Phê duyệt chiến lược quốc gia quản lý tổng hợp chất thải rắn đến năm 2025 tầm nhìn đến năm 2050* (Decision 2149/2009/QD-TTg by the Prime Minister on approval on national strategy on integrated solid waste management until 2025 vision toward 2050). Hanoi: Prime Minister (PM). https://www.uncrd.or.jp/content/documents/PM%20 Decision%20of%20Approval%20of%20NSISWM%20(Eng).pdf.   Accessed 24 Apr 2020.

Takagi, H., Ty, T. V., Thao, N. D., & Esteban, M. (2014). Ocean tides and the influence of sea-level rise on floods in urban areas of the Mekong Delta. *Journal of Flood Risk Management, 8,* 292–300.

Tanner, T., Bahadur, A., & Moench, M. (2017). *Challenges for resilience policy and practice* (Working paper 519). Overseas Development Institute.

UN Water. (2020). *World water development report 2020.* https://en.unesco.org/ themes/water-security/wwap/wwdr/2020. Accessed 31 Mar 2020.

Van Den Berg, K., & Duong, T. C. (2018). *Solid and industrial hazardous waste management assessment: Options and actions areas (English).* Washington, DC: World Bank Group. http://documents.worldbank.org/curated/en/352371563196189492/Solid-and-industrial-hazardous-waste-management-assessment-options-and-actions-areas. Accessed 24 Apr 2020.

Van den Ham, L., Janssen, J., & Van der Meulen, L. (2019). Internship report: Circular Park Can Tho, waste management and catcher studies, VHL UAS.

WWF. (2009). *The greater Mekong and climate change: Biodiversity, ecosystem services and development at risk* (p. 34). Bangkok: WWF.

CHAPTER 7

Climate Resilience in Urban Informal Settlements: Towards a Transformative Upgrading Agenda

Matthew French, Alexei Trundle, Inga Korte, and Camari Koto

Abstract Informal settlements are on the frontline in the battle against climate change. Home to one billion people, their infrastructure deprivations pose challenges for the health and resilience of communities and ecosystems. Upgrading of informal settlements can improve urban services and infrastructure, strengthen tenure security, and empower local

M. French (✉)
Monash Sustainable Development Institute, Monash University,
Melbourne, VIC, Australia
e-mail: matthew.french@monash.edu

A. Trundle
Connected Cities Lab/Melbourne Sustainable Society Institute, University of
Melbourne, Parkville, VIC, Australia

I. Korte
United Nations Human Settlements Program (UN-Habitat), Suva, Fiji

C. Koto
College of the Marshall Islands, Uliga, Majuro Atoll, Marshall Islands

© The Author(s) 2021
R. de Graaf-van Dinther (ed.), *Climate Resilient Urban Areas*,
Palgrave Studies in Climate Resilient Societies,
https://doi.org/10.1007/978-3-030-57537-3_7

communities. This chapter examines the conceptual and practice relationships between climate resilience and in-situ upgrading. It critiques prevailing approaches, which centre upon threshold, coping, recovery, and adaptive capacities. Transformative capacity offers greater scope for addressing climate change impacts at a level commensurate with the size of the challenge, and for redressing the entrenched structural inequalities and deep socio-spatial injustices shaping cities in the Global South that perpetuate vulnerability and socio-spatial exclusion. Five elements are identified to advance transformative informal settlement upgrading: socio-technical innovation; a climate justice framing; greater attention to intersectional dimensions; inclusive governance and community empowerment; and fit for purpose finance.

Keywords Informal settlements • Urban poor • Climate resilience • Slum upgrading • Climate change • Housing

7.1   INTRODUCTION

Approximately one billion people in the Global South live in urban informal settlements, many of which are highly vulnerable to the impacts of climate change. Informal settlements, also referred to as slums or squatter settlements, generally develop outside formal regulatory and planning systems. They are defined as being groups of dwellings that lack one or more of the following: access to improved water; access to improved sanitation; sufficient living area; durability of housing; and security of land tenure[1] (UN-Habitat, 2003). Occupation and construction take place simultaneously, over many years or decades. These processes are driven by residents who take it upon themselves to meet their housing needs in the absence of formal alternatives that are affordable, accessible, or proximal to socio-economic opportunities and services (Acioly and French 2012).

As a persistent modality of urban and housing development, informal settlements are symptomatic of structural inequalities and the limitations

---

[1] The fifth measure—tenure security—is excluded from global estimates due to a lack of global agreement on the appropriate measure of this term (UN-Habitat 2016). As such, figures of one billion urban dwellers living in informal settlement conditions under-represent these characteristics globally.

of formal land and housing supply. Although globally the proportion of urban dwellers living in informal conditions fell significantly between 2000 and 2015 (from 39 to 30 percent), the total number of inhabitants of informal settlements increased by almost 100 million (UNDESA 2015:40). With the addition of a further 1.1 billion city dwellers expected between 2015 and 2030, these areas will almost certainly grow in total population over the next decade and beyond (ibid).

Urban informal settlements pose significant environmental challenges for people and ecosystems. Rates of infections and diarrhoeal disease are high due to poor water and sanitation services (Lilford 2017). Informal settlements have higher rates of disease burden and premature mortality than formal areas (Sverdlik 2011). Key environmental determinants of health in informal settlements include spatial segregation, insecure residential status, housing structural quality, water and sanitation, and climate change risks (Corburn and Sverdlik 2019). Women and children are often the most severely impacted by unhealthy informal conditions (Chant and McIlwaine 2016). Ecosystems and the eco-servicing of cities are compromised by the incremental, unplanned encroachment of informal settlements, a process which in turn increases the vulnerability of cities to environmental stressors and shocks such as extreme weather events (Zari et al. 2019). Health pandemics, such as COVID-19, exacerbate the risks and vulnerabilities for informal settlement populations (French et al. 2020).

'In-situ upgrading' is a common approach to improve informal settlement access to urban services and infrastructure, while also strengthening tenure security and empowering local communities (UN-Habitat 2015). This approach contrasts that of eviction and the replacement of slums with new housing, which are widely recognised as having detrimental impacts on the livelihoods and social fabric of residents, while also incurring prohibitive financial and political costs for the institutions that deploy them (Pugh 2001). It is within this normative framework and praxis of 'in-situ upgrading' that the climate resilience agenda and responses are manifested, particularly when focused through the lens of sub-system vulnerabilities in the cities of the Global South.

7.2   CLIMATE RESILIENCE IN URBAN
INFORMAL SETTLEMENTS

The urban poor that informally occupy coastal and riverine areas stand to suffer some of the greatest impacts of climate change. Because of their limited access to formal services, utilities, and infrastructure, informal settlements are generally more sensitive than formal areas to climate-related shocks and stresses, and face increased exposure to an array of extreme and slow-onset climate events (Kabisch et al. 2015). As outlined in Table 7.1, the tendency for informal settlements to occupy otherwise un-developed 'at-risk' areas such as intertidal zones and floodplains means that sea level rise and coastal inundation have heightened implications for the livelihoods, health, and wellbeing of their occupants. Increasing rainfall variability and intensity can also exacerbate flooding and waterlogging in the face of negligible drainage and sanitation infrastructure. The increasing frequency of extreme weather events also compounds existing socio-spatial vulnerabilities and strains coping and relief systems (Bulkeley and Tuts 2013; Friend and Moench 2013; Kabisch et al. 2015; Trundle et al. 2019).

The relationship between vulnerability and resilience is contested. In Holling's seminal writings (1973) on the concept, resilience was observed and defined as a measure of the persistence and absorbance of *instability*, a characteristic that is integral to the generation of resilience in ecological systems over time. Miller et al. (2010) argue that resilience and vulnerability vary in conceptualisation and theory, methodologies, and practice, despite their common consideration of shocks and stressors. However, if resilience is to be applied normatively to socio-centric systems such as cities, a fundamental social consideration is the redistributive consequences that resilience practice entails. Vulnerability provides a critical entry point into these wider ethical considerations that are often left unconsidered in resilience practice (Olsson et al. 2015).

Climate resilience has become an increasingly prevalent planning framework for urban governance, design, and development, including in the Global South. Within the 'resilience turn' in urban policy (Meerow and Stults 2016), 'urban resilience' and 'climate resilience' have emerged as two of the most frequently referenced contemporary modes of resilience thinking over the last decade (Trundle 2020). In both instances, respective considerations of urban subject systems and anthropogenically modified shocks and stresses diverge from earlier ecological and engineering modes

Table 7.1 Climate resilience issues and projected impacts in urban informal settlements

Climate change hazards	City-scale vulnerability	Additional informal vulnerability factors
Higher (and increasing) average temperatures	General exposure to extreme heat and related heat stress, particularly for at-risk populations	Worsened localised heat impacts due to compact, poor quality built form, and a lack of utilities
Increasing rainfall variability and intensity, including extreme rainfall events	Stormwater infrastructure and water supply capacities limited to existing variability. Recurrent transport and services disruptions, localised damage	Impacts exacerbated due to a lack of regulatory compliance and formal drainage infrastructure. Proximity to informal water sources (rivers, springs, ocean) heightening exposure
Potable water quality and availability	Risk of utility shortfalls in supply, potable water importation needs, price spikes, and expensive infrastructure investment	Limited income capacity to pay for potable water price spikes, dependencies on variable informal water sources with additional risk of contamination
Sea level rise and coastal inundation (including extreme events, e.g. storm surges)	Demand for enhanced coastal defences, loss of previously insurable urban areas, long-term relocation strategies in conflict with local short-term interests and market structures	Limited options for retreat and reinforcement due to spatial and financial constraints, as well as a lack of state resourcing/subsidies. Tendency to be located in already at-risk areas, including intertidal zones. Lack of sanitation leading to secondary risks (e.g. disease)
Increasing frequency of more intense tropical cyclones/ hurricanes	Building standards superseded by new return periods and intensities, heightened demand for state evacuation facilities, disruptions to urban food supplies	Direct dependency on localised food production with limited capacity to enter the cash economy. Poor quality/non-compliant housing construction. Frequent exclusion from evacuation and relief supply considerations
Ecosystem services deterioration	Increasing demand for imported goods and pressure on peri-urban surrounds	Widespread dependency on ecosystem services for primary or secondary income and subsistence
Climate-induced migration	Rapid urban growth risks outpacing development planning and government capacities	Heightened pressure on informal areas outside of government housing and land release programmes

Adapted from: UN-Habitat (2018), Trundle et al. (2019), Moser and Satterthwaite (2010)

of resilience thinking (Holling 1996). Specifically, their focus on subjectively defined systems, and alternative transformations in system states, creates a normative resilience 'vision' that is open to contestation by constituent social groups and actors (Olsson et al. 2015).

Uptake of the resilience agenda in the Global South has been both triggered by climate-related shocks and through development practice (Leitner et al. 2018). For example, the Rockefeller Foundation's *Asian Cities Climate Change Resilience Network (ACCCRN)* engaged ten Asian cities to build the 'adaptive capacity of vulnerable urban populations' (da Silva et al. 2012 p.127). Globally, within the *2030 Agenda for Sustainable Development,* the 11th 'urban' Sustainable Development Goal aims to make cities 'inclusive, safe, resilient and sustainable' and includes a number of indicators and targets on resilient upgrading informal settlements (Griggs et al. 2013:16).

7.3   In-Situ Settlement Upgrading

The contemporary upgrading model evolved from a trajectory of approaches over the past century. Informal settlements that proliferated in developing countries in the early twentieth century were seen in a negative light and government policy focused on eviction and eradication (Abrams 1966). By mid-century, large-scale, high-density housing projects aimed to replace slums with modernist housing; however, they could not meet demand and informal settlements continued to expand (Pugh 2001). A paradigm shift occurred in the late-1960s. Emphasis was given to the potential of the urban poor to meet their shelter needs through self-help, providing them the 'freedom to build' (Turner 1972). Meanwhile, informal settlements continued to expand due to supply, affordability, and location constraints of the 'sites and services' approach (Wakely 2018; Abbott 2002).

By the end of the twentieth century, in-situ (on site) upgrading of informal settlements had become mainstream. Informal communities were increasingly recognised as agents in upgrading decision-making, especially women and girls (Massey 2017). Community involvement improved project design, implementation, and increased the sustainability of interventions (Patel et al. 2001; French et al. 2019; Pugh 2001). In parallel, there was increased awareness of the importance of integrating urban planning and land management for the purposes of socio-spatial integration of informal settlements in their wider urban systems (French and Lalande 2013; UN-Habitat 2015).

Contemporary upgrading programmes include a standard suite of components (Table 7.2). Improving access to safe water and sanitation is often

Table 7.2 Key components of in-situ upgrading and their relationship to climate resilience

Component	Underlying logic and objectives	Examples	Relationship to climate resilience
Data and socio-spatial inclusion	Inclusion of settlements in urban plans and governance Political recognition of residents and rights Identify local needs and deficiencies	Community-based enumerations Inclusion in government census Citizen science Spatial mapping of existing conditions and vulnerabilities	Accurate data upon which to design interventions and respond in crises Capacity and awareness of climate hazards
Community participation	Plan with, not for residents Build social capital Ensure diverse voices are heard Ensure residents' needs and priorities are reflected in upgrading actions Smooth implementation and promote ownership of interventions 'Right to the city'	Form and/or strengthen existing community committees and groups Capacity building of civil society groups Set-up and/or strengthen governance forums for formal exchange and dialogue with authorities	Build social resilience through collective action Identify and implement community-led adaptation actions Empower and provide space for the urban poor to engage in city and national resilience governance processes Strengthen the social contract between citizens and the state
Social support	Strengthen existing social capital Improve livelihoods Reduce conflict, crime and exclusion	Youth and women's groups Micro-finance and community-led savings groups	Build social resilience and adaptive capacity Foster local ownership and action
Land tenure regularisation	Regularise settlement layout Increase security of tenure to foster a sense of ownership and inclusion Promote endogenous investment in housing improvements Increase local land-based revenues	Issue certificates of land occupancy (household and/or community) Issue land titles Form community land trusts Street addressing and house numbering	Limit further informal land encroachment on hazardous land Strengthen the asset base of urban poor Identification and safeguarding of public land for climate-responsive ecosystem services

(continued)

Table 7.2   (continued)

Component	Underlying logic and objectives	Examples	Relationship to climate resilience
Sanitation and water supply	Improve human health, wellbeing, and dignity Reduce burden of ill-health, especially on children and vulnerable groups Improve productivity and reduced burden of poor water and sanitation, especially for women and girls	Centralised or decentralised wastewater treatment Water supply through reticulated systems, wells, or others Community-Led Total Sanitation (CLTS); faecal sludge removal; Water, Sanitation, and Hygiene (WASH)	Diversify water sources to strengthen resilience to droughts or extreme weather events Reduce infrastructure operating vulnerability to climate change impacts
Urban services (i.e. electricity, transportation, solid waste collection, IT)	Extend urban services to improve amenity, mobility, and opportunity Formalise services that are currently informal and/or illegal Improve public health	Extend formal electricity and information technology networks Municipal solid waste collection Bus stops, cable cars	Strengthened resilience to adapt to climate change and severe weather events; better connection to formal services during extreme climatic events
Drainage and environmental improvements	Reduce ecosystem and environmental contamination Reduce vulnerability to environmental effects such as flooding and storm surges Improve public health	Stormwater drainage Green spaces and tree planting Flood mitigation infrastructure Wetland and natural waterway restoration Relocation of households occupying vulnerable land	Significant links between environmental improvements and strengthened climate resilience Improved precinct and citywide resilience

(*continued*)

Table 7.2 (continued)

Component	Underlying logic and objectives	Examples	Relationship to climate resilience
Street, public space, and green infrastructure	Improve accessibility for residents and emergency services Improve amenity, safety, and security Reduce spatial segregation through integration with neighbouring areas Stimulate local economic development	Paving roads, streets, and pathways Installing street lighting Establishment or upgrading of public spaces such as sports areas, plazas, parks Public facilities such as libraries, service hubs	Reduced vulnerability to extreme weather events Emergency access in times of crises Increased green space reducing urban heat island effect
Housing rehabilitation	Improve housing quality and amenity Reduce density and overcrowding	Promote incremental housing improvements Improved house services (e.g. clean cookstoves)	Green building design to reduce environmental impact Improved built environment resilience to shocks and hazards

Adapted from: UN-Habitat (2014, 2015, 2018), Zevenbergen et al. (2015), Lucci et al. (2015), Cities Alliance (2020)

prioritised, having direct outcomes for resident health, wellbeing, and dignity. The paving of streets and laneways, along with urban 'acupuncture' in the form of public space interventions, is also widely deployed in tandem with stormwater management and drainage. Formalisation of services such as electricity, street lighting, and solid waste collection is also commonplace. Enhancement of tenure security through partial recognition and documentation of land or occupancy rights is also used to reduce the threat of eviction, exploitation, and corruption. This legitimisation also promotes endogenous local investment and provides the potential for leveraging local land-based revenues.

Climate resilience is increasingly being considered as part of upgrading processes to promote their integration with their local ecological and environmental conditions and accommodate projected climate change impacts (Zevenbergen et al. 2015; Satterthwaite et al. 2020). These additional considerations demand a '*citywide*' approach to ensure that concepts such as vulnerability and climate change projections are applied consistently and efficiently (UN-Habitat 2015). The citywide approach spatially documents all informal settlements within a city, understands the nature of spatial and socio-economic segregation and exposure to disasters and environmental risks, develops a citywide strategy/plan for upgrading, and establishes and implements local financial and institutional structures for delivery over an extended programme period.[2]

There is considerable variation in the arrangement of upgrading components across regions, countries, and cities. Satterthwaite et al. (2020) note upgrading projects can be seen as a continuum: a 'ladder', from 'upgrading that is actually eviction' to 'transformative upgrading' which is rooted in deep partnerships between authorities and communities where integrated, comprehensive, environmentally responsive upgrading occurs. There is significant variation across the world. Citywide upgrading is a standard part of government policy in Latin America and the Caribbean (Magalhães 2016). In contrast, upgrading has yet to become mainstream in Africa due to incipient urban and population growth, fiscal constraints, and the scale of urban informality (Bah et al. 2018; Gulyani and Bassett 2007). Community-led, urban acupuncture projects are common in Asia (e.g. Boonyabancha 2009). The Pacific presents unique conditions and responses given the interconnectivity between a diverse array of socio-cultural, spatial, and temporal urban migration patterns and the spatial typologies and climate vulnerabilities faced by Small Island Developing States (Jones 2016; McEvoy et al. 2020).

7.4   Climate Resilience Capacities in Informal Settlement Upgrading

The physical infrastructure components of slum upgrading can be understood as an attempt to increase the *threshold capacity* of settlement residents and their environments. Physical interventions, as outlined in Table 7.2, can help reduce vulnerability to the impacts of more intense precipitation, flooding, sea level rise, and coastal erosion, and can also

---

[2] The second important element of a citywide approach is increasing the supply of formal housing concurrently with citywide upgrading (Payne 2005). This recognises that to reduce the formation of new informal settlements, and to provide housing for informal settlement households who may be affected by upgrading interventions, the supply of new housing which is affordable to low-income households is an essential part of housing and upgrading policy. Brazil's 'My House, My Life' programme is a notable example: UN-Habitat (2013) *Scaling-up affordable housing supply in Brazil: The 'My House, My Life' program.* United Nations Human Settlements Program: Nairobi.

reduce their vulnerability to the long-term effects of climate change such as increased water scarcity, urban heat island effects, and disease vectors. Likewise, intangible upgrading components, as outlined in Table 7.2, can also be important levers to increase *threshold capacity*. Land regularisation and tenure security, data and mapping of settlements, and community participation and citizen engagement can be fundamental components of strengthening climate resilience.

The underlying assumption embedded in current upgrading policy and practice is that investments in urban infrastructure (*threshold capacity*) contribute to improved *coping capacity* and *recovery capacity*. When designed, implemented, and managed well, physical upgrading interventions can help to reduce damage during extreme weather events in informal settlements, and help households and informal communities to recover more effectively. Land tenure regularisation is an important factor in increasing the coping and recovery capacity of households (Unger et al. 2017). Likewise, actions to increase the social and institutional threshold capacity of communities can pay dividends for coping during disasters and extreme weather events, and for recovering from them, especially when processes and response structures build on the ingenuity and creativity of residents and where humanitarian assistance builds on ongoing urban resilience efforts rather than undermines them.

While *adaptive capacity* is increasingly considered in settlement upgrading, it has yet to become mainstream. It is operationalised to ensure that investments in the physical upgrading of streets, stormwater drainage, and water and sanitation will be sustainable in the face of medium- and long-term environmental change, notably rising sea levels, increased ambient temperature, precipitation, and more frequent extreme weather events. Climate resilient technological interventions are increasingly being considered, for example rainwater harvesting to diversity water sources, on-site sanitation systems that are less vulnerable to trunk infrastructure outages, and natural filtration of stormwater to reduce contamination of ecosystems (Brown et al. 2018). However, these are not commonplace in upgrading practice, and there remains very little leapfrogging to climate-adaptive approaches.

In coastal areas, the most significant adaptive capacity demand is to respond to rising sea levels and flooding. In the face of these interconnected threats and associated extreme weather events, 'managed retreat – the relocation of homes and infrastructure under threat from coastal flooding – is one of the few policy options available' (Tadgell et al. 2018:102). This is exacerbated by the limited funding available for in-situ adaptive upgrading relative to the scale of and continued growth in informality in the Global South. Relocation, however, is immensely problematic. Johnson (2020) argues that relocation is a flawed response and often equates to forced eviction. Relocation is expensive; Governments in the Global South have insufficient fiscal capacity to provide safe alternatives commensurate with demand. When alternative housing can be provided, it is often poorly located vis-a-vis livelihood opportunities; does not match the needs, priorities, and fiscal capacity of the intended beneficiaries; and the relocation disrupts existing social support systems.[3] Relocation projects are seldom framed in terms of long-term development action for affected communities, but rather short-term risk reduction measures which limit their effectiveness (Johnson 2020). Low-income households and communities have considerable formal and informal existing assets that can be harnessed for endogenous adaptive responses (Moser and Satterthwaite 2010), yet such an asset-based framework is often ignored to justify relocation. Overall, because relocation is so complex and returns policy and practice to a previously failed approach to addressing informal settlements, relocation as an adaptive action is not commonplace.

7.5   Transformative Informal Capacities: Elements of a Holistic Approach to Urban Climate Resilience

The first four capacities—threshold, coping, recovery, and adaptive—are alone unlikely to meet the scale of climate challenges faced by coastal informal settlements. Instead, the resilience of urban areas to the impacts of climate change hinges most critically on building the fifth 'transformational capacity' proposed by Ovink (Ovink, H. Personal communication, 22 February 2019). This conceptual division provides a useful tool to explore how a step change in policy and practice responses to informal settlements might be advanced. Many of the elements of transformative capacity discussed in Table 1.1 in Chap. 1 are relevant for informal settlements in the Global South. For example, proactive and inclusive planning and design with all stakeholders, linking water issues to urban dynamics, can provide a framework for moving from single projects to integrating innovative processes through the enhancement and participatory inclusion of local capacities.

---

[3] While there are unique cases of relatively successful relocation (e.g. Cronin and Guthrie 2011), these are the exceptions rather than the norm.

In addition, there are specificities related to the Global South context that deserve detailed interrogation and critical reflection. These include extreme poverty, socio-economic exclusion, food insecurity, protracted conflict, entrenched gender inequalities, nebulous land and housing rights, corruption, weak governance and institutions, and, importantly, the rapid pace of urbanisation and population growth (Pieterse 2011). As Friend and Moench observe, these urban systems 'are emergent mosaics … which reflect social values and relations, coupled with the coevolving environmental and infrastructure systems' (2015, p. 646). High level of informality is a dominant feature of these mosaics that cannot simply be ignored, overridden, or undermined by formal approaches. Therefore, framing climate resilience as an optional 'add-on' to informal settlement upgrading in the Global South is unlikely to produce the transformative capacity that is needed at a city scale.

We argue that five elements are needed to underpin climate-resilient transformative capacity building that integrates the informal domain. These five elements are not exhaustive, nor are they independent of each other. They build on informal settlement upgrading theory, policy, and practice with a critical view towards a greater ambition to achieve transformation.

The first element is ***socio-technical innovation.*** Physical upgrading projects too often rely on conventional approaches to engineering infrastructure and services, for example centralised trunk infrastructure systems for water and sanitation provision. These require significant capital investment, have high embodied energy, and require fossil fuel for their operation and maintenance. Innovative twenty-first century technologies and infrastructure approaches can help 'leapfrog' to more sustainable urban infrastructures that can better respond to local conditions and work with nature and local environmental systems and processes to deliver services in harmony with the environment and climatic changes. Infrastructure innovations can reduce the 'lock-in' to environmentally intensive infrastructures and, when well designed, can be more resilient to local shocks (Brown et al. 2018). Importantly, it demands a *socio*-technical approach that reframes the relationship between social systems and technical engineering components. A socio-technical approach focuses on the interaction of people and their environment, for example environmental exposure pathways affecting human health in informal settlements, to achieve mutual benefit for people as well as the environment.[4]

---

[4] The Revitalising Informal Settlements and their Environments (RISE) programme is an example of leapfrogging informal settlement water and sanitation infrastructure to more environmentally sustainable socio-technical approaches. See: www.rise-program.org and Brown et al. (2018).

Linked with socio-technical innovation is *climate justice*, the second element for building transformative capacities. By reframing climate change from a predominantly environmental or physical phenomenon towards one that is inherently social, ethical, and political, the severe inequalities and human rights violations that underpin socio-spatial exclusion and vulnerability are brought to the fore. Too often climate resilience policy and practice 'lacks a normative focus on advancing the needs of the most marginalised and most vulnerable' (Bartlett and Satterthwaite 2016:25). Likewise, practice-driven considerations of resilience through donor-developed frameworks and mechanisms do not always engage sufficiently with the theoretical principles and the term's conceptual strengths (Sharifi 2016). An explicit climate justice framing can give due recognition that disadvantaged groups, such as those occupying coastal informal settlements, are on the frontlines of the fight against climate change impacts and suffer grave inequalities and political and socio-economic exclusion that exacerbates their vulnerability. Ziervogel et al., for example, provide four 'entry points' for a justice- and rights-based resilience agenda, which stem from the recognition of the endogenous resilience qualities and capacities that can exist outside of—and be in conflict with—the institutional regime-level functions that are empowered at a city-scale (2017).

Third, a climate justice framing encourages greater attention be paid to the *intersectional dimensions* and diversity of lived experiences of informal settlement residents and groups. Intersectionality is a framework for understanding how combinations of vulnerabilities, injustices, and discrimination (i.e. gender, class, race, sexuality, age, ability, poverty, etc.) are manifested and experienced differently by different groups 'due to their situatedness in power structures based on context-specific and dynamic social categorisations' (Kaijser and Kronsell 2014; Grünenfelder and Schurr 2015). Evidence across a variety of contexts shows that women, girls, sexual minorities, and less able people living in informal settlements suffer specific and compounding marginalities and vulnerabilities, and suffer disproportionately from the effects of poor informal settlement services and conditions (Castán Broto and Neves Alves 2018). A climate justice and intersectional framing therefore demands 'a drastic shift in existing power structures and the policies these structures bring forward, a shift that puts the needs of the poor high on the agenda' (Roy et al. 2016:6).

Fourth, **inclusive governance and community empowerment** are crucial underpinnings of transformational capacity. Informal settlement residents and organised community groups should be meaningfully empowered to play a role in the decisions affecting their lives and neighbourhoods. The enormous social capital that exists in informal settlements cannot be underestimated and can be a powerful force for transformation (Satterthwaite et al. 2020; Archer 2016). Community empowerment alone, however, is insufficient for transformation, and a significant strengthening of urban governance is needed. Urban governance institutions in the Global South are poorly resourced, overburdened, and have limited capacity to inclusively plan and manage urban development processes. Aid and donor investments can further undermine the endogenous capacity and functioning of state and city authorities, especially in countries highly dependent on foreign aid (French et al. 2019).

A more flexible framing of urban governance and community engagement as they relate to climate resilience is advantageous. Harris et al. (2018) argue that too often the climate resilience agenda is premised on notions of 'win-win' end-state situations rather than seen as highly political, complex, and dynamic processes (Mikulewicz 2019). Coined *'negotiated resilience'*, 'by foregrounding the procedures and processes of resilience, we can better attend to the politics and stakes of negotiation — i.e., whose interests are advanced in what way and with what possible outcomes, as well as how ideals of consensus or policy agendas are actively sought, managed, and at times produced' (Harris et al. 2018:196). Helmke and Levitsky's typology of informal–formal interactions provides a useful framework for considering the compatibility of informal and formal functional values relative to the existing effectiveness of urban governance (2004). Applications to informal settlements focused on endogenous climate resilience capacities highlight the importance of such an approach, drawing out areas where informality is providing supportive substitutive urban functions, and others where competing values require negotiation (Trundle 2020).

In practice, this means asking capacity and resilience *for whom* and *for what*, as well as *where*, *when*, and *why* (Meerow and Newell 2019). This is particularly important in the Global South given the complexity of decision-making, with the allocation of scarce resilience resources competing against demands for poverty reduction (Friend and Moench 2013), the prevalence of corruption and elite capture (Berquist et al. 2015), and the additional agendas and conditions attached to large volumes of sustainable urban development funding (Mikulewicz 2019). These considerations are widely recognised in urban development literature as structurally constraining the capacity to respond in humanitarian crises and socio-spatial injustices stemming from historic and contemporary urban growth patterns (Dodman et al. 2013; Borie et al. 2019). However, they have had less attention within the conceptual frameworks deployed in the practice of urban climate resilience.

The fifth element is **fit-for-purpose finance**. This recognises that achieving a transformation in informal settlement upgrading requires significantly increased levels of finance compared with previous and current levels, and requires new financial modalities and instruments underpinned by climate justice framing (UN-Habitat 2018). Transformation efforts should amplify the plethora of 'bottom-up', community-led financing initiatives for settlement upgrading projects that have proliferated over the past two decades (Archer 2012). Community-based finance is premised on the power of collective self-help and support at the local level, and seeing finance as an essential part of building the transformative capacity of the urban poor.[5]

While community-based finance is an important mechanism, it has had limited success in accessing international climate finance (i.e. the Green Climate Fund (GCF) and Global Environment Facility (GEF)). Colenbrander et al. (2018:902) argue that 'adaptation finance is primarily allocated to multilateral entities and national governments, rather than local organizations. This means that the social, political and economic processes that create and sustain inequalities within a country will be the same processes that determine how adaptation finance is used'.[6] Similarly, Overseas Development Assistance (ODA) and well-intentioned philanthropy-backed initiatives are too-often premised on short-term, project-based action which cannot foster transformational change for climate resilience, especially at city and community levels (Ayers 2009). 'Fit-for-purpose' financing, therefore, must include expanded access to climate finance by local governments (e.g. municipalities) and community groups to fund climate resilience processes, not projects (Funder et al. 2015) (Table 7.3).

---

[5] For example, community-development funds (CDFs) initiated by the Asian Coalition for Community Action provide funding to informal settlement community groups across more than 100 cities (Archer 2012). Similarly, the Community-Led Infrastructure Finance Facility (CLIFF) provides capacity grants and revolving capital funds to non-profit and community groups to make investments in housing and upgrading (McLeod and Mullard 2006; World Habitat 2020).

[6] For example, based on empirical case study research in Malawi, Barrett (2014) demonstrates a stark mismatch; the areas most in need received relatively little finance, and therefore the 'distribution of adaptation funds do not support the larger goal of climate justice'.

Table 7.3 Assessment of prevailing upgrading components and climate resilience measures against the five capacities framework

Components of upgrading	Threshold capacity (prepare and prevent damage resulting from environmental variation)	Coping capacity (reduce damage during extreme weather events)	Recovery capacity (recover effectively after disasters)	Adaptive capacity (adapt to current and expected environmental trends)	Transformative capacity (to transition proactively to a climate-resilient society)
Data and socio-spatial inclusion	Baseline and vulnerability assessments; Data informs preparations and prevention actions	Authorities know about settlements and their populations and can respond	Pre-disaster/ event data available for recovery and rebuilding	Data can inform upgrading projects to adapt to anticipated trends	Data could be used for (citywide) transformation, but often is not
Community participation and social support	Communities mobilised to prepare and prevent damage; City plans incorporate the urban poor; early-warning systems	Crisis plans; Community support networks respond; mutual self-help	Community support networks to recover; humanitarian assistance; mutual self-help	Communities often have more immediate needs that become focus of today's efforts	Values misalignment; Urban poor not included in citywide risk resilience governance processes
Land tenure regularisation	Land regularisation can foster/increase mitigation investments	Land tenure clarity can reduce conflict during events	Strengthened land rights enables faster recovery; reduced conflict	Land regularisation in hazardous sites in conflict with long-term adaptation;	Clear land rights is a prerequisite for transformation, but not seen through a climate lens

(continued)

Table 7.3 (continued)

Components of upgrading	Threshold capacity (*prepare and prevent* damage resulting from environmental variation)	Coping capacity (*reduce* damage during extreme weather events)	Recovery capacity (*recover effectively* after disasters)	Adaptive capacity (*adapt* to current and expected environmental trends)	Transformative capacity (to *transition proactively* to a climate-resilient society)
Sanitation and water supply	Increased diversity of water sources; reduced environmental contamination	Reduced vulnerability to public health impacts	Quicker and safer recovery; Water, Sanitation and Hygiene (WASH) projects	Diversity and resilience water and sanitation possible but not commonplace; little leapfrogging	Interventions are largely delivered using old methods; little 'leapfrogging'
Street and public space infrastructure; Drainage and environmental improvement; Urban services	Improves access and can minimise climate change and disaster impacts; Greatly strengthens urban resilience	Settlements with upgraded physical infrastructure far better placed to cope during	If built well, provides a good foundation to rebuild and recover	Innovative adaptive approaches possible but not commonplace; often exacerbate potential impacts	Limited. Actions rely on traditional approaches
Housing improvements	Limited action in current upgrading efforts	Limited action in current upgrading efforts	'Build back better' agenda. Limited recovery effort for housing - mostly occupant-led.	Very limited practical action in urban informal settlements	Very limited practical action in urban informal settlements

Source: Author's compilation

7.6 Conclusion

This chapter used the 'five capacities' framework as a tool to examine climate resilience of in-situ informal settlement upgrading to draw novel theoretical and conceptual insights. It contributes to a critique of prevailing climate resilience approaches that centre on simplistic *threshold*, *coping*, *recovery*, and *adaptive* capacities. These appear insufficient to make a meaningful and significant change given the scale of the challenge with one billion people living in informal settlements today, and the fact that these first four capacities cannot overcome the entrenched structural inequalities and deep socio-spatial injustices shaping cities in the Global South.

The fifth capacity, *transformative capacity*, offers an opportune pathway that may foster more significant material improvements in the human health of informal settlement residents and mitigate the impacts of climate change at a level commensurate with the scale and urgency of the challenge. For transformative capacity, the chapter identified five transformational elements for informal settlement upgrading theory and praxis: socio-technical innovation; a climate justice framing; greater attention to intersectional dimensions; inclusive governance and community empowerment; and fit for purpose finance. This is reflective of a broader need for climate change responses and projects to reform the prevailing models of urban development in the Global South, which perpetuate structural inequalities and unsustainable urban development practices.

This chapter aligns with other critiques of the normative framing of resilience, particularly in dynamic and contested cities of the Global South. Central to these critiques is a lack of theoretical and practical mechanisms for understanding and addressing divergent normative viewpoints and values (Davoudi et al. 2012). As outlined, we have argued that any such consideration implicitly requires a more explicit framing in terms of fundamental social attributes such as equity and justice (Chelleri et al. 2015; Ziervogel et al. 2017; Trundle 2020). By operationalising resilience through a climate justice lens and building on existing approaches to upgrading practice, we demonstrate that informal settlements, exemplars of urban subsystems with divergent functions and values, could be transformed as part of efforts to generate climate resilient urban areas in coastal and delta cities.

REFERENCES

Abbott, M. (2002). An analysis of informal settlement upgrading and critique of existing methodological approaches. *Habitat International, 26*(3), 303–315.

Abrams, C. (1966). *Squatter settlements: The problem and the opportunity.* Washington, DC: Department of Housing and Urban Development.

Acioly, C., & French, M. (2012). *Housing developers: Developing world. International encyclopedia of housing and home.* p. 422–428, London: Elsevier.

Archer, D. (2012). Finance as the key to unlocking community potential: Savings, funds and the ACCA programme. *Environment and Urbanization, 24*(2), 423–440. https://doi.org/10.1177/0956247812449235.

Archer, D. (2016). Building urban climate resilience through community-driven approaches to development: Experiences from Asia. *International Journal of Climate Change Strategies and Management, 8*(5), 654–669. https://doi.org/10.1108/IJCCSM-03-2014-0035.

Ayers, J. (2009). International funding to support urban adaptation to climate change. *Environment and Urbanization, 21*(1), 225–240. https://doi.org/10.1177/0956247809103021.

Bah, E. M., Faye, I., Geh, Z. F. (2018). Slum Upgrading and Housing Alternatives for the Poor. In: *Housing Market Dynamics in Africa.* Palgrave Macmillan, London.

Barrett, S. (2014). Subnational climate justice? Adaptation finance distribution and climate vulnerability. *World Development, 58,* 130–142. https://doi.org/10.1016/j.worlddev.2014.01.014.

Bartlett, S., & Satterthwaite, D (Eds.) (2016). *Cities on a finite planet: towards transformative responses to climate change.* Earthscan: London.

Berquist, M., Daniere, A., & Drummond, L. (2015). Planning for global environmental change in Bangkok's informal settlements. *Journal of Environmental Planning and Management, 58*(10), 1711–1730, https://doi.org/10.108 0/09640568.2014.945995.

Boonyabancha, S. (2009). Land for housing the poor – By the poor: Experiences from the Baan Mankong nationwide slum upgrading programme in Thailand. *Environment and Urbanization, 21*(2), 309–329. https://doi.org/10.1177/0956247809342180.

Borie, M., Pelling, M., Ziervogel, G., & Hyams, K. (2019). Mapping narratives of urban resilience in the global south. *Global Environmental Change, 54,* 203–213. https://doi.org/10.1016/j.gloenvcha.2019.01.001.

Brown, R. R., Leder, K. S., Wong, T., French, M., Ramirez, D., Chown, S. L., Luby, S., Clasen, T., Reidpath, D., El Sioufi, M., McCarthy, D. T., Forbes, A. B., Simpson, J., Allotey, P., & Cahan, B. (2018). Improving human and environmental health in urban informal settlements: the Revitalising Informal Settlements and their Environments (RISE) programme. *The Lancet Planetary Health, 2*(29), 29–29. https://doi.org/10.1016/S2542-5196(18)30114-1.

Bulkeley, H., & Tuts, R. (2013). Understanding urban vulnerability, adaptation and resilience in the context of climate change. *Local Environment, 18*(6), 646–662. https://doi.org/10.1080/13549839.2013.788479.

Castán Broto, V., & Neves Alves, S. (2018). Intersectionality challenges for the co-production of urban services: Notes for a theoretical and methodological agenda. *Environment and Urbanization, 30*(2), 367–386. https://doi.org/10.1177/0956247818790208.

Chant, S., & McIlwaine, C. (2016). *Cities, slums and gender in the global south: Towards a feminised urban future.* London: Routledge. ISBN 9780415721646.

Chelleri, L., Waters, J., Olazabal, M., & Minucci, G. (2015). Resilience trade-offs: Addressing multiple scales and temporal aspects of urban resilience. *Environment and Urbanization, 27*(1), 181–198. https://doi.org/10.1177/0956247814550780.

Cities Alliance. (2020). *A Policy Framework for a Slum Upgrading Program. Webpage.* Retrieved April 15, 2020. https://www.citiesalliance.org/policy-framework-for-slum-upgrading-programme.

Colenbrander, S., Dodman, D., & Mitlin, D. (2018). Using climate finance to advance climate justice: The politics and practice of channelling resources to the local level. *Climate Policy, 18*(7), 902–915. https://doi.org/10.1080/14693062.2017.1388212.

Cronin, V., & Guthrie, G. (2011). Community-led resettlement: From a flood-affected slum to a new society in Pune, India. *Environmental Hazards, 10*(3–4), 310–326, https://doi.org/10.1080/17477891.2011.594495.

Corburn, J., & Sverdlik, A. (2019). Informal Settlements and Human Health. In: Nieuwenhuijsen M., Khreis H. (eds.), *Integrating Human Health into Urban and Transport Planning.* Springer

Dodman, D., Brown, D., Fracis, K., Hardoy, J., Johnson, C., & Satterthwaite, D. (2013). Understanding the nature and scale of urban risk in low- and middle-income countries and its implications for humanitarian preparedness, planning and response. *IIED working paper.* London, March 2013.

da Silva, J., Kernaghan, S., Luque, A., da Silva, J., Kernaghan, S., & Luque, A. (2012). A systems approach to meeting the challenges of urban climate change. *International Journal of Urban Sustainable Development, 4*(2), 125–145. https://doi.org/10.1080/19463138.2012.718279.

Davoudi, S., Keith, S., Jamila Haider, L., Quinlan, A. E., Peterson, G. D., Wilkinson, C., Fünfgeld, H., McEvoy, D., & Porter, L. (2012). Resilience: A bridging concept or a dead end? 'Reframing' resilience: Challenges for planning theory and practice. *Planning Theory and Practice, 13*(2), 299–333. https://doi.org/10.1080/14649357.2012.677124.

De Graaf-van Dinther, R. E., & Ovink, H. (2020). The five pillars of climate resilience. In R. E. De Graaf-van Dinther, *Climate resilient urban areas. Technology, governance and development in coastal delta cities.* Palgrave Macmillan: London, UK.

French, M. A., & Lalande, C. (2013). Green Cities Require Green Housing: Advancing the Economic and Environmental Sustainability of Housing and Slum Upgrading in Cities in Developing Countries. In: Simpson R., Zimmermann M. (eds.), *The Economy of Green Cities. Local Sustainability*, vol 3. Springer, Dordrecht. https://doi.org/10.1007/978-94-007-1969-9_24.

French, M., Popal, A., Rahimi, H., Popuri, S., & Turkstra, J. (2019). Institutionalizing participatory slum upgrading: A case study of urban co-production from Afghanistan, 2002–2016. *Environment and Urbanization, 31*(1), 209–230. https://doi.org/10.1177/0956247818791043.

French, M., Ramirez-Lovering, D., Sinharoy, S., Turagabeci, A., Ihsan, L., Leder, K., & Brown, R. (2020). *Informal settlements in a COVID-19 world: moving beyond upgrading and envisioning revitalisation, Cities & Health.* https://doi.org/10.1080/23748834.2020.1812331.

Friend, R., & Moench, M. (2013). What is the purpose of urban climate resilience? Implications for addressing poverty and vulnerability. *Urban Climate, 6*, 98–113. https://doi.org/10.1016/j.uclim.2013.09.002. Elsevier.

Funder, M., Christoplos, I., Friis-Hansen, E., Lindegaard, L., & Pain, A. (2015). *Making the green climate fund work for the poor.* Policy brief. Danish Institute for International Studies. https://www.files.ethz.ch/isn/189588/pb_green_climate_fund_web.pdf

Griggs, D. et al. (2013). Sustainable development goals for people and planet. *Nature, 495*(7441), 305–7. https://doi.org/10.1038/495305a.

Grünenfelder, J. & Schurr, C. (2015). Intersectionality – A challenge for development research and practice. *Development in Practice, 25*(6), 771–784, https://doi.org/10.1080/09614524.2015.1059800.

Gulyani, S., & Bassett, E. M. (2007). Retrieving the baby from the bathwater: Slum upgrading in sub-Saharan Africa. *Environment and Planning C: Government and Policy, 25*(4), 486–515. https://doi.org/10.1068/c4p.

Harris, L. M., Chu, E. K., & Ziervogel, G. (2018). Negotiated resilience. *Resilience, 6*(3), 196–214. https://doi.org/10.1080/21693293.2017.1353196.

Holling, C. S. (1973). Resilience and stability of ecological systems. *Annual Review of Ecology and Systematics. Annual Reviews, 4*(1), 1–23. https://doi.org/10.1146/annurev.es.04.110173.000245.

Holling, C. S. (1996). Engineering resilience versus ecological resilience. *Engineering Within Ecological Constraints, 31*, 32. https://doi.org/10.17226/4919.

Johnson, C. (2020). The implications of climate related resettlement policies in cities of the global south. *Planning Theory & Practice, 21*(1), 125–154. https://doi.org/10.1080/14649357.2020.1704130.

Jones, P. (2016). Informal urbanism as a product of socio-cultural expression: Insights from the Island Pacific. In S. Attia, S. Shabka, Z. Shafik, & A. Ibrahim (Eds.), *Dynamics and resilience of informal areas*. Cham: Springer.

Kabisch, S., Jean-Baptiste, N., John, R., & Kombe, W. J. (2015). Assessing Social Vulnerability of Households and Communities in Flood Prone Urban Areas, In Pauleit, S. and et al. (eds.), *Urban Vulnerability and Climate Change in Africa. Future Cit*, pp. 287–318. https://doi.org/10.1007/978-3-319-03982-4.

Kaijser, A., & Kronsell, A. (2014). Climate change through the lens of intersectionality. *Environmental Politics, 23*(3), 417–433. https://doi.org/10.108 0/09644016.2013.835203.

Leitner, H., Sheppard, E., Webber, S., & Colven, E. (2018). Globalizing urban resilience. *Urban Geography, 39*(8), 1276–1284. https://doi.org/10.108 0/02723638.2018.1446870.

Lilford, R. et al. (2017). Improving the health and welfare of people who live in slums. Series, the health of people who live in slums. *The Lancet, 389*(10068), 559–570. https://doi.org/10.1016/S0140-6736(16)31848-7.

Lucci, P., Bhatkal, T., Khan, A, & Berliner, T. (2015). *What works in improving the living conditions of slum dwellers – a review of the evidence across four programmes*. Overseas Development Institute: London. https://www.odi.org/sites/odi.org.uk/files/odi-assets/publications-opinion-files/10188.pdf.

Magalhães, F. (2016). *Slum upgrading and housing in Latin America*. Washington, DC: Inter-American Development Bank.

Massey, R. T. (2017). The effect of informal settlement upgrading on Women's social networks: Layout versus location. *Urban Forum, 28*, 205–217. https://doi.org/10.1007/s12132-017-9302-7.

McEvoy, D., Mitchell, D., & Trundle, A. (2020). Land tenure and urban climate resilience in the South Pacific. *Climate and Development, 12*(1), 1–11. https://doi.org/10.1080/17565529.2019.1594666.

McLeod, R., & Mullard, K. (Eds). (2006). *Bridging the Finance Gap in Housing and Infrastructure*. ITDG: Rugby.

Meerow, S., & Newell, J. (2019). Urban resilience for whom, what, when, where, and why? *Urban Geography, 40*(3), 309–329. https://doi.org/10.108 0/02723638.2016.1206395.

Meerow, S., & Stults, M. (2016). Comparing conceptualizations of urban climate resilience in theory and practice. *Sustainability, 8*(701), 1–16. https://doi.org/10.3390/su8070701.

Mikulewicz, M. (2019). Thwarting adaptation's potential? A critique of resilience and climate-resilient development. *Geoforum*. https://doi.org/10.1016/j.geoforum.2019.05.010.

Miller, F., et al. (2010). Resilience and vulnerability: Complementary or conflicting concepts? *Ecology and Society, 15*(2), 1–11. https://doi.org/10.5751/ES-03378-150311.

Moser, C., & Satterthwaite, D. (2010). *Toward pro-poor adaptation to climate change in the urban centres of low- and middle-income countries.* Washington, DC: World Bank.

Olsson, L., et al. (2015). Why resilience is unappealing to social science: Theoretical and empirical investigations of the scientific use of resilience. *Science Advances, 1*(4), 1–11. https://doi.org/10.1126/sciadv.1400217.

Patel, S., Burra, S., & D'Cruz, C. (2001). Slum/Shack Dwellers International (SDI) – foundations to treetops. *Environment and Urbanization, 13*(2), 45–59. https://doi.org/10.1177/095624780101300204.

Payne, G. (2005). Getting Ahead of the Game: A Twin-Track Approach to Improving Existing Slums and Reducing the Need for Future Slums. *Environment and Urbanization, 17*(1), 135–145.

Pieterse, E. (2011). Recasting urban sustainability in the South. *Development, 54,* 309–316. https://doi.org/10.1057/dev.2011.62.

Pugh, C. (2001). The theory and practice of housing sector development for developing countries, 1950–99. *Housing Studies, 16*(4), 399–423.

Roy, M., Cawood, S., Hordijk, M., & Hulme, D. (2016). *Urban poverty and climate change: Life in the slums of Asia, Africa and Latin America.* London: Routledge.

Satterthwaite, D., Archer, D., Colenbrander, S., Dodman, D., Hardoy, J., Mitlin, D., & Patel, S. (2020). Building resilience to climate change in informal settlements. *One Earth.* https://doi.org/10.1016/j.oneear.2020.02.002.

Sharifi, A. (2016). A critical review of selected tools for assessing community resilience. *Ecological Indicators, 69,* 629–647. https://doi.org/10.1016/j.ecolind.2016.05.023. Elsevier.

Sverdlik A . 2011. Ill-health and poverty: a literature review on health in informal settlements. *Environment and Urbanization, 23*(1), 123–155. https://doi.org/10.1177/0956247811398604.

Trundle, A. (2020). Resilient cities in a Sea of Islands: Informality and climate change in the South Pacific. *Cities, 97.* https://doi.org/10.1016/j.cities.2019.102496. Elsevier.

Trundle, A., Barth, B., & McEvoy, D. (2019). Leveraging endogenous climate resilience: Urban adaptation in Pacific Small Island developing states. *Environment and Urbanization, 31*(1), 53–74. https://doi.org/10.1177/0956247818816654. Sage.

Turner, J. (1972). Housing as a verb. In J. Turner & R. Fichter (Eds.), *Freedom to build: dweller control of the housing process.* London: Macmillan.

UNDESA. (2015). *World Urbanisation Prospects: 2015.* UNDESA: New York.

UN-Habitat. (2003). *The challenge of slums: Global report on human settlements 2003.* United Nations Human Settlements Programme. London: Earthscan.

UN-Habitat. (2014). *Streets as tools for urban transformation in slums.* UN-Habitat: Nairobi

UN-Habitat. (2015). *A practical guide to designing planning and executing city-wide slum upgrading programmes.* Nairobi: UN-Habitat.

UN-Habitat. (2018). *Addressing the most vulnerable first: Pro-poor climate action in informal settlements.* Nairobi: UN-Habitat.

Unger, E.-M., Zevenbergen, J., & Bennett, R. (2017). On the need for pro-poor land administration in disaster risk management, *Survey Review, 49*(357), 437–448, https://doi.org/10.1080/00396265.2016.1212160.

Wakely, P. (2018). *Housing in developing cities: experience and lessons learned.* Routledge: London.

World Habitat Awards. (2020). Website. Accessed 10 March 2020. https://world-habitat.org/world-habitat-awards/winners-and-finalists/community-led-infrastructure-finance-facility-cliff/.

Zari, M. P., Kiddle, G. L., Blaschke, P., Gawler, S., & Loubser, D. (2019). Utilising Nature-Based Solutions to Increase Resilience in Pacific Ocean Cities. *Ecosystem Services, 38*(June). https://doi.org/10.1016/j.ecoser.2019.100968.

Zevenbergen, C., Olthuis, K., Benni, J., Eichwede, K. (2015). Slum Upgrading: Assessing the importance of location and a plea for a spatial approach. *Habitat International, 50*, 270–288. https://doi.org/10.1016/j.habitatint.2015.08.033.

Ziervogel, G., Pelling, M., Cartwright, A., Chu, E., Deshpande, T., Harris, L., Hyams, K., et al. (2017). Inserting rights and justice into urban resilience: A focus on everyday risk. *Environment and Urbanization, 29*(1), 123–138. https://doi.org/10.1177/0956247816686905.

A Transformative Process for Urban Climate Resilience: The Case of Water as Leverage Resilient Cities Asia in Semarang, Indonesia

Naim Laeni, Henk Ovink, Tim Busscher,
Wiwandari Handayani, and Margo van den Brink

Abstract The Water as Leverage Resilient Cities Asia program introduced a transformative process to explore, design and implement inclusive, integrated and holistic climate solutions for three cities in Asia. Focusing on the city of Semarang, Indonesia, this chapter aims to assess whether, and if

N. Laeni (✉) • T. Busscher • M. van den Brink
Department of Spatial Planning and Environment, Faculty of Spatial Sciences,
University of Groningen, Groningen, The Netherlands
e-mail: n.laeni@rug.nl

H. Ovink
Special Envoy for International Water Affairs at Kingdom of the Netherlands,
The Hague, The Netherlands

W. Handayani
Department of Urban and Regional Planning, Faculty of Engineering,
Diponegoro University, Semarang, Indonesia

© The Author(s) 2021
R. de Graaf-van Dinther (ed.), *Climate Resilient Urban Areas,*
Palgrave Studies in Climate Resilient Societies,
https://doi.org/10.1007/978-3-030-57537-3_8

so, how the program has been able to build resilience capacities and to draw lessons that can be applied in other cities. Our analysis shows that the program in Semarang developed an enabling environment and has increased transformative capacity by building strong stakeholder coalitions between multidisciplinary design teams, governmental agencies, knowledge partners and NGOs. The collaboration between these actors led to the development of an innovative programmatic approach, which comprises six interrelated, inclusive and integrated climate solutions. When implemented, these solutions will also increase the city's threshold, coping and adaptive capacity. The implementation of such solutions, however, proves to be a complex financial and institutional challenge.

Keywords Water as Leverage • Capacity building • Transformative process • Collaborative planning process • Semarang

8.1   INTRODUCTION

Asian coastal cities are known for being prone to climate risks and water-related disasters. Not only are climate impacts, such as heat waves, rising sea levels and changing rainfall patterns, expected to have severe economic, environmental and societal consequences, Asian coastal cities are also typically among the highest-ranked cities worldwide in terms of exposed population (see, e.g., Pillai et al. 2010; PBL Netherlands Environmental Agency 2018). The United Nations World Urbanization Prospects has estimated that by 2030 the urban Asian population will constitute half of the global urban population (United Nations 2015). This means that the urban Asian population will account for 83% of the world's population at risk due to rising sea levels (Vinke et al. 2017).

The Water as Leverage Resilient Cities Asia program (WaL), a Dutch government initiative, conducted a climate vulnerability assessment which identified Asia as a region at risk. Their findings reconfirmed that South Asian and Southeast Asian cities are indeed extremely vulnerable to climate impacts (Dutch Special Envoy for International Water Affairs 2019). At the same time, the results also presented possibilities to strengthen the climate adaptation capacity of the region, such as by stimulating sustainable finance for proactive and integrated climate solutions. The WaL program currently runs in three cities: Khulna in Bangladesh, Chennai in

India and Semarang in Indonesia. In each of these cities, experts and stakeholders have been brought together to analyze the context-specific situation and risks and to collaboratively develop and design integrated climate solutions, which are eventually to be scaled-up and replicated in other cities in the respective regions (Government of the Netherlands 2019a).

This chapter focuses on the WaL program in the city of Semarang. Semarang is one of the rapidly growing metropolitan cities in Indonesia. The city center of Semarang is located in a low-lying area between mountains and the coastline and is severely pressured by sea-level rise and urban growth. Due to continuous urbanization and the development of weak clay-layer sediment, the city of Semarang is sinking. The annual subsidence rate averages at 6 to 7 centimeters with extremes of 14 to 19 centimeters (Mulyana et al. 2013). This urbanization has also significantly increased the amount of impermeable paving and decreased the number of green and open areas, which in turn exacerbated water runoff problems and reduced the amount of rain that percolates into the soil. These have led to recurring annual flash floods during the tropical season.

Traditionally, the Semarang government's flood risk management response has mainly focused on conventional and reactive 'hard' infrastructural measures such as the normalization of rivers, building (temporary) structures including dams and walls, and the installation of water pumps in various canals and rivers. A sole focus on infrastructure measures is, however, becoming increasingly considered insufficient, since this response is seen as constraining the options for adapting to current and future urban and climate risks. Instead, the call for developing holistic climate solutions and building climate adaptation capacity is becoming louder (Handayani et al. 2019).

To answer to this call, Semarang has become part of various global policy platforms and networks, which are aimed at improving urban climate resilience. For instance, Semarang has participated in the 100 Resilient Cities program, pioneered by the Rockefeller Foundation, and has developed a City Resilience Strategy (CRS). The city is also part of the City Resilience Program of the World Bank. These programs provide the city with resources and knowledge support and exchange, which facilitate the city to develop and implement resilience programs and strategies. Semarang, therefore, is experienced in building coalitions and collaborating with international partners in developing urban climate solutions. The WaL program in Semarang builds on these experiences as it aims to improve the adaptive capacity of the city so that it can cope with current and future social and environmental challenges from climate-related impacts. The program therefore introduced a so-called 'transformative process', which included an analysis of the context as well as an exploration, design and implementation of more inclusive, integrated and holistic climate resilience solutions.

The aim of this chapter is to assess whether, and if so, how the WaL program in Semarang has been able to build resilience capacities and to draw lessons that can be applied in other cities. In doing so, we first explain the program activities that have been undertaken during the different phases of the WaL program lifecycle in Semarang. This is followed by a section in which we analyze and reflect on how and to what extent the program was able to build resilience capacities (De Graaf 2009) as part of its transformative process. In the final section, we present our conclusions and suggestions for future research.

8.2   Water as Leverage Resilient Cities Asia

The Water as Leverage Resilient Cities Asia program (WaL) is an international initiative that focuses on establishing an enabling environment for the development, implementation and replication of innovative holistic climate-related solutions for Asian cities through a collaborative approach and shared design process (Government of the Netherlands 2019a). The program was initiated in 2018 by the Dutch Special Envoy for International Water Affairs in partnership with Dutch Ministries,[1] the Dutch Development Bank (FMO) and various international partners.[2] In support of the collaborative and inclusive process, the program provides funding for marshaling relevant experts and stakeholders at the city, regional, national and international levels. The process enabled the stakeholders to develop urban climate adaptation strategies, actions (projects) and financial arrangements (see Fig. 8.1). In particular, Water as Leverage has actively stimulated the engagement of financial actors in the early phases of the project lifecycle (Government of the Netherlands 2019a).

---

[1] Ministry of Foreign Affairs, Ministry of Infrastructure and Water, Netherlands Enterprise Agency.

[2] IABR, AWB, AIIB, GCA, 100RC, FMO, WB, ADB, GCF, IsDB, UNHabitat, WWF, Pegasys, PfR, OECD.

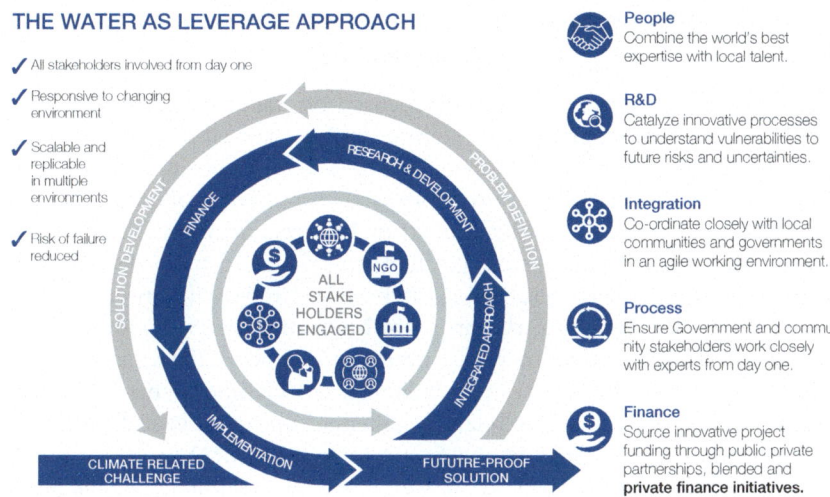

Fig. 8.1   The Water as Leverage Approach for climate resilient solutions. (Source: Dutch Special Envoy for International Water Affairs 2019)

The WaL program encompasses a transformative approach for climate adaptation and climate finance. The approach aims to address three common problems (Government of the Netherlands 2019a). Firstly, the limited participation of diverse actors across all phases of the project lifecycle, from understanding to implementation, was addressed through the introduction of an inclusive and collaborative process. The inclusion of a wide range of actors, such as citizens, local knowledge institutions and financial parties, should have also helped to overcome fragmentation and to shift the focus away from only short-term project gains. Secondly, the transformative approach aimed to bridge the gap in funding mechanisms for multidisciplinary and collaborative research and development. Finally, through its focus on proactive, integrated and holistic climate solutions, the approach presented opportunities for meeting the targets of the 2030 Agenda for Sustainable Development (UN SDGs). The remainder of this chapter will explain the four different phases of the WaL program.

8.2.1   Phase 1: Understanding and Exploring Vulnerabilities and Opportunities

The first phase of the WaL program focused on gaining a deeper understanding of climate-related vulnerabilities in Asia, as well as exploring the opportunities to enhance local capacities to develop innovative urban climate solutions. In this phase, environmental and urban challenges related to water and climate risks in the context of Asia were analyzed, hotspots of global climate challenges were identified, and the need for proactive and holistic solutions was highlighted (Dutch Special Envoy for International Affairs 2019). In doing so, the program built on research conducted by The Netherlands Environmental Enterprise Agency (2018) known as 'The Geography of Future Water Challenges',[3] which performed a global scenario study to assess water-related challenges up to the year 2050. The WaL program report 'Setting the Scene for A Call for Action' outlined that, firstly, the region faces great water and environmental challenges such as water shortage, high salinity levels, pollution, land subsidence and frequent and intensive flooding (Dutch Special Envoy for International Affairs 2019). Secondly, the report identified almost 30 Asian climate vulnerable cities considering their diverse urban and water challenges in relation to both extreme events and long-term developments. Based on this report, the program selected three cities in terms of their environmental challenges with opportunities for developing innovative urban climate solutions (see Fig. 8.2). The selection of the three cities, which were Chennai in India, Khulna in Bangladesh and Semarang in Indonesia, was also based on the presence of existing water and climate coalitions and the willingness to collaborate with international partners on water management and climate resilience programs. On a national level, Bangladesh, India and Indonesia are part of the Global Commission on Adaptation that seeks to accelerate the implementation of concrete solutions to address climate challenges through the exchange of knowledge and international cooperation. On a city level, Semarang and Chennai are both part of the 100 Resilient Cities Network,[4] which is an international platform that provides resources and opportunities to exchange and collaborate with other cities to develop and implement resilience strategies and initiatives (Handayani et al. 2019).

---

[3] An initiative of the Dutch Special Envoy for International Water Affairs.

[4] 100 Resilient Cities transitioned into the Global Resilient Cities Network and the Resilient Cities Catalyst in 2019.

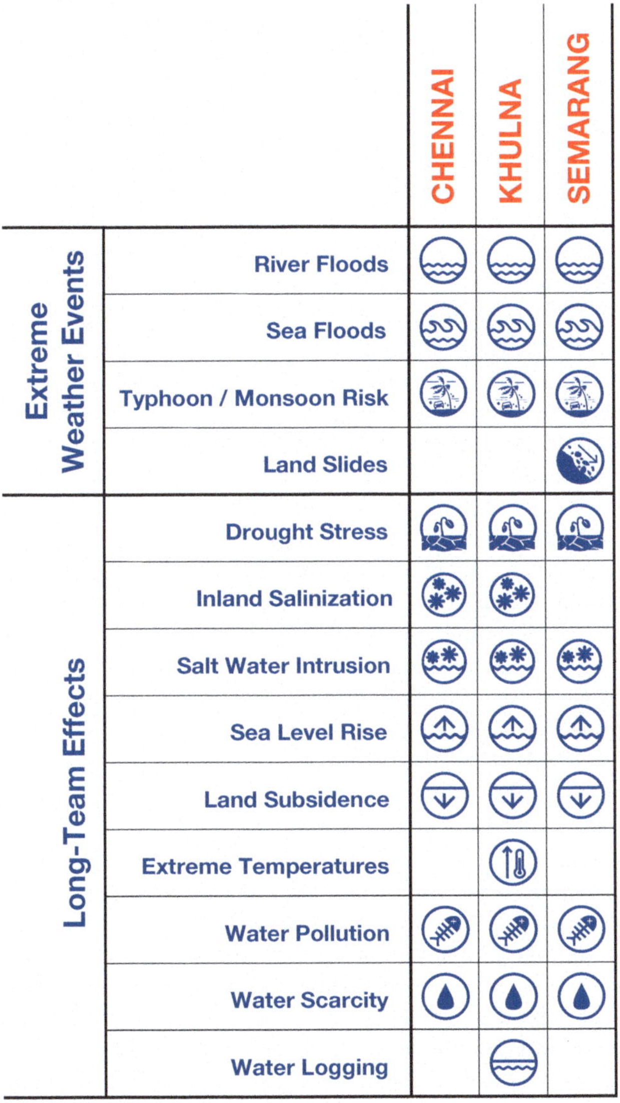

		CHENNAI	KHULNA	SEMARANG
Extreme Weather Events	**River Floods**	〰	〰	〰
	Sea Floods	🌊	🌊	🌊
	Typhoon / Monsoon Risk	🌴	🌴	🌴
	Land Slides			🏔
Long-Team Effects	**Drought Stress**	🌵	🌵	🌵
	Inland Salinization	✳	✳	
	Salt Water Intrusion	✳	✳	✳
	Sea Level Rise	⬆	⬆	⬆
	Land Subsidence	⬇	⬇	⬇
	Extreme Temperatures		🌡	
	Water Pollution	🐟	🐟	🐟
	Water Scarcity	💧	💧	💧
	Water Logging		⊜	

Fig. 8.2 Assessment of the water challenges for the city selection. (Source: Water as Leverage Program, Setting the Scene for A Call for Action)

8.2.2   Phase 2: Creating and Strengthening an Enabling Environment

The second phase of the WaL program revolved around the Call for Action and the activities of the multidisciplinary design teams. Each multidisciplinary design team was tasked with building strong local coalitions within the city between the city governments, knowledge partners and networks of NGOs. Collectively, these local coalitions had to conduct in-depth research on the local context and develop conceptual designs for potential climate solutions. The Call for Action was launched on Earth Day 2018 (22nd April). Initially, almost 40 multidisciplinary design teams applied. However, only six teams were selected, or two for each city (Government of the Netherlands 2019a). The assessment criteria for selecting the teams were divided into three sets. The first set was related to the problem definition, including the understanding of the water system, climate urgencies and the urban context of the city. The second set concerned the programs, such as the extent to which these aligned with the UN SDG goals and the UN Valuing Water Principles. The third set was related to the composition of the teams, which considered gender, age balance and the combination of internationally and locally based organizations (Government of the Netherlands 2019a). The two multidisciplinary design teams selected in Semarang were 'One Resilient Semarang' and 'Cascading Semarang'. The former consisted of One Architecture & Urbanism, Deltares, Wetlands International, Kota Kita, Sherwood Design Engineers, Hysteria Grobak, Iqbal Reza and UNDIP. The latter included MLA+, Deltares, FABRICations, PT Witteveen+Bos Indonesia, UNDIP, UNISSULA and IDN Liveable Cities (Water as Leverage 2019).

Initially, the two multidisciplinary design teams each conducted research and analysis separately and designed their climate resilience programs for Semarang (Cascading Semarang 2018; One Resilient Semarang 2018). In support of these activities, WaL organized a series of local, regional and international workshops to involve relevant actors from public authorities, financial and academic institutions, and local communities in the research and analysis stages for designing the climate solutions. This introduced the teams and their coalitions to marginalized communities, responsible government agencies and a network of national and international financial institutions. This helped to build both system knowledge and produce inclusive and integrated climate solutions that could deliver multiple benefits for Semarang. In other words, these workshops helped

Fig. 8.3 Conceptual design of the WaL strategic climate resilience programs in Semarang. (Source: Water as Leverage 2019 – image produced by the two teams One Resilient Semarang and Cascading Semarang)

to enhance an understanding of the context and to co-create strategic resilience programs (Government of the Netherlands 2019b) (see also Fig. 8.3). Interestingly, over the course of time, the two teams started to work together, eventually resulting in one combined strategic climate resilience program for Semarang, which consisted of six innovative climate resilience programs (Water as Leverage 2019). The six strategic programs are described in Table 8.1, and include solutions that focus on nature-based approaches, integrated spatial design, inclusive development and long-term resilience (Water as Leverage 2019). The programs were presented to and discussed with the Mayor of Semarang, representatives from the national government and other stakeholders in March 2019, during the Water as Leverage International Seminar in Semarang (Water as Leverage 2019).

8.2.3 Phase 3: Brokering Financial Arrangements for Project Implementation

Water as Leverage is currently in the third phase of the program lifecycle. Central to this phase is identifying and brokering sustainable climate

Table 8.1 Strategic climate resilience programs in the Water as Leverage program in Semarang (Water as Leverage 2019)

Strategic programs	Responsible teams	Measures
1. Water-neutral industry (demand)	One resilient Semarang	This program aims to promote value-adding sustainable industrial development through the revitalization of existing industrial clusters and their surrounding urban areas, applying water-neutral strategies, and using innovative technologies and services. The measures are focused on using less water, optimizing water supply systems, building storage and conveyance system, and reusing and recycling water.
2. Feeding industry (supply)	Cascading Semarang	The main goal of this program is to shift the water consumption paradigm in Semarang by reducing or stopping groundwater extraction in the industrial sector. The program proposes alternative ways for increasing industrial water supply through capturing, conveying, storing, treating and reusing water runoff.
3. Network of resilience Kampungs	One resilient Semarang	This program aims to introduce a decentralized approach for green and blue infrastructure investment to stimulate the transformative capacity at the community level. The program further aims to strengthen bottom-up governance modes by encouraging community participation in planning and budgeting processes. The program thus intends to enable a sense of ecological citizenship, participatory water planning, stewardship of shared resources, and accelerate infrastructure improvement for the city. The measures include physical interventions, capacity building and resource allocation for climate vulnerable communities.
4. Integrated protective coastal zone	One resilient Semarang	This program proposes an integrated vision and plan for a protective and productive coastal zone. The program aims to ensure coastal protection through a combination of hard infrastructure with soft (mangrove) foreshores. This should enhance long-term industrial competitiveness, urban and rural development, and biodiversity. The program includes land use planning measures such as creating a long-term vision for integrated coastal protection and the concentrating and clustering of high-value industrial areas.

(*continued*)

Table 8.1 (continued)

Strategic programs	Responsible teams	Measures
5. Spongy Mountain terraces	Cascading Semarang	This program proposes the integration of spatial measures with flood management in hilly areas. This should result in an increase of the city's adaptive capacity. The program aims to increase the resilience of existing and upcoming urban developments by working on a watershed scale. The proposed measures are designed to create spaces for water and to increase land stabilization through increasing green areas, shallowing aquifers for rainwater harvesting and creating green terraces for slope stabilization. These measures should enable flexible ways of living with water.
6. Rechanneling the city	Cascading Semarang	This program focuses on increasing the threshold capacity of the city center's urban water system. It should improve urban water management, create additional capacity for water storage and regulate water flow. The program includes several measures to improve urban water systems and reinforce the existing water networks, such as by improving public space along canals, formalizing (or revitalizing) informal settlement communities along canals, increasing wetland areas, and widening and deepening rivers and canals.

funding for the implementation of the strategic climate resilience programs. Two key processes can be identified through which WaL attempts to broker and stimulate cooperation between the multidisciplinary design teams, knowledge partners, the city government and (inter)national financial institutions (Government of the Netherlands 2019a). The first process revolves around the organization of financial workshops in which international financial institutions such as the Dutch Development Bank (FMO), the Asian Infrastructure Investment Bank, the World Bank, the Asian Development Bank, the Islamic Development Bank and the Green Climate Fund participate. The Regional Workshop in Singapore in April 2019, for instance, provided opportunities for the multidisciplinary teams to present their programs, discuss the financial criteria for climate adaptation finance and explore potential funding opportunities. Simultaneously, these workshops were also seen as an important activity for the exchange of lessons learned between cities, thereby increasing institutional capacities among partners and local stakeholders (Government of the Netherlands 2019a).

The second process involves bilateral cooperation between the Netherlands and participating countries under the existing Memorandum of Understanding on water cooperation. The Netherlands, in this way, stimulates international cooperation between the national governments and each city to support implementation of the strategic climate resilience program. The detailed financial programs to receive funding for project implementation are now being developed per project and per city. To be able to receive funding from international financial institutions and from government budgets, the strategic climate resilience programs need to be approved by national and local governments and incorporated in national and local planning documents. This is an arduous and complex task as it requires the involvement of the national government from an early stage, a shared strategy from the city government and the design teams on the promotion of the strategic programs, and the formalization of these programs into the national, regional and local development plans. To coordinate this process and to engage with governmental agencies at the national level in Jakarta, the WaL program, together with local Semarang stakeholders and the multidisciplinary team, formed a 'Water as Leverage Taskforce'. During the time of writing this chapter (April 2020), it was still unclear whether the program would be successful in acquiring governmental approval and funding and whether subsequent financial arrangements with financial institutions could be made.

8.2.4   Phase 4: Replication and Scaling-up

The fourth and final phase of the WaL program focuses on replication and scaling-up. The three selected cities are expected to serve as exemplary pilots for bringing sustainable solutions and transformative capacity to the agenda of similar cities in Asia and other parts of the world. Ideally, the innovations at both the project and program level (i.e., the transformative process and the innovative climate solutions) would be scaled up to be applied internationally (Dutch Special Envoy for International Water Affairs 2019). At the level of the proposed projects, WaL aimed to identify financial arrangements and implementation opportunities (see also Fig. 8.1) to replicate climate solutions in other cities and regions. At the program level, WaL introduced a new approach and process in Asia, emphasizing inclusive and collaborative pre-project development, the development of holistic and integrated design solutions, and continuous engagement with local, provincial, regional and national governments and international financial partners (Government of the Netherlands 2019a). This includes innovative ways to fund climate resilience solutions and is a two-way process. Firstly, spatial designers in charge of developing proactive and integrated climate solutions need to become aware of bankability issues and incorporate this understanding into the design process and the designs. Secondly, international financial institutions should acknowledge the increased risk tolerance of financing innovative climate resilience programs and (help) develop mechanisms to validate and evaluate holistic and integrated design programs. In addition, replication and scaling-up could also involve local stakeholders such as the city government, universities and knowledge partners who are involved in other local and international resilience programs. These actors can safeguard the continuity of the program and continue the learning process since these parties will also share their experience and insights with other cities and countries.

8.3   WATER AS LEVERAGE: TOWARD TRANSFORMATIVE ADAPTATION?

Climate change adaptation involves long-term systemic change. Studies show that enabling such change requires a clear break away from incremental, or business-as-usual solutions, such as sectorial-based and reactive measures. Instead, transformative adaptation (Few et al. 2017) and transformative capacity need to be built for adapting to current and long-term challenges. The WaL program has tried to achieve this through numerous methods. The program has expanded the scope and the number of disciplines involved in the development of climate solutions. As a result, different measures are proposed, which were less focused on building dams and flood walls, river normalization and drainage improvements. Moreover, the program has initiated collaborative planning processes such as community empowerment programs and co-creative design processes that build upon systematic analysis, exploration and understanding of the local context (Water as Leverage 2019). Overall, the following key lessons can be drawn from the WaL program in Semarang.

8.3.1   Reflection on the Co-creation of Program-Based Urban Climate Solutions

1. *System knowledge and local learning process for urban climate adaptation*

The first key lesson is that the WaL program has been able to formulate a comprehensive and inclusive, yet feasible, problem definition, upon which holistic and context-sensitive urban climate solutions could be developed. The first phase of the program, namely the research and exploration phase, has increased system knowledge and established an overall understanding of the water system, urban dynamics and the social context. The second phase of the program has encouraged the multidisciplinary teams to conduct a deeper analysis of the context and opportunities for interventions. In this process, local communities, private actors and other relevant local stakeholders were also invited to participate and to learn together in understanding the problems and co-creating the solutions. In addition, through the development of the six strategic programs, the program has provided a learning platform that has facilitated the understanding that water is not a sectoral problem that can be solved with mere traditional infrastructure works. In other words, the program has provided a platform for local capacity building to ensure that transformation will occur with the support of local governments, local NGOs, academics, and community representatives.

2. *Development of an innovative and programmatic approach*

The second lesson is the development of an innovative programmatic approach for Semarang. Combined, the six different strategic programs create a holistic and integrated program that is clearly aimed at increasing Semarang's threshold as a result of improving its coping and adaptive capacity (De Graaf 2009) for urban climate resilience (see Table 8.1), as outlined in Chap. 1. Firstly, the strategic programs for increasing the industrial water supply, namely the 'Water-neutral industry' and 'Feeding industry', aim to divert the extensive use of groundwater resources to other alternatives, thereby anticipating and reducing future consequences from land subsidence. These programs contribute to the threshold capacity of the city by increasing water supplies and diversifying sources of water supplies. Secondly, the flood-proof modes of urbanization presented in the strategic programs 'Spongy Mountain terraces' and 'Rechanneling the city' involve experimentation with flexible measures and with spatial adaptation to reduce urban flood risks. Both programs contribute to the adaptive capacity of the city. The 'Integrated Protective Coastal Zone' program is an example of climate-sensitive spatial planning, by for instance restoring natural wetland areas for the community's adaptation. Third and finally, the 'Network of Resilience Kampungs' program aims to ensure sustainable community involvement with a particular focus on the needs of vulnerable people with regard to water supply and energy, as well as on the coping capacity for flood response and mitigation.

8.3.2 Reflection on the Financing and Implementation of the Urban Climate Solutions

3. *Creating an enabling environment and building capacities for implementation*

The third lesson relates to the development of an enabling environment by the WaL program through which the program has increased the opportunity for successful project development and implementation, resembling what Allmendinger and Haughton (2009) describe as 'soft spaces' for planning. As Kaczmarek (2018: 182) explains, these are 'areas where deliberate attempts are made to introduce new and innovative ways of thinking, especially in places where there is considerable resistance to cross-sectoral and inter-territorial governance'.

During the different program phases and through involving various governmental levels (from local to international), sectors and disciplines, the WaL program facilitated various informal focus group discussions and a series of informal meetings and visits. These involved local governments, local NGOs, academics, international financial institutions and community representatives to help understand and co-create the climate resilient solutions. Facilitating these different forms of collaboration enabled coalition building and the mutual understanding of the potentials and challenges of the different actors involved, which is a prerequisite for long-term implementation. Consequently, by creating an enabling environment for inclusive collaboration and building transdisciplinary networks, the program has increased the transformative capacity of Semarang, as defined in Chap. 1 of this book.

The emerging combination of the various roles between the different levels of government showcased how WaL's bottom-up approach challenged the existing structures, and how the existing mechanisms of top-down policy challenged the development and implementation of the WaL program. Although the program facilitated an innovative bottom-up approach, it may not be able to operationalize their programs and projects 'smoothly' due to the limited authority and capacity of the local government. Decentralization policy in the country has not yet led to its expected outcomes, as at the practical level bottom-up initiatives without sufficient top-down support will not function effectively in the current governance system in Indonesia. Transformative adaptation requires coalitions that are, on the one hand, supported by the national government—since the national government is the main financier of the strategic programs—and, on the other hand, entail robust coordination at the ground level to guarantee long-term commitment from local stakeholders to ensure sustainable implementation.

4. *Securing integrated and sustainable climate finance*

The fourth lesson concerns the WaL program's attempts to secure or broker integrated and sustainable financial investments. Sustainable financing of the strategic programs is essential because it allows the implementation of more measures to cope with climate challenges in the long run. This is a critical and challenging issue in Indonesia as well as in other places around the world, as the local governments involved in the program only have limited financial budgets. Besides asking for support from the national government, many of these local governments lack the capacity to explore and access alternative funding opportunities. As a consequence, the national government has always played an important and dominant role in the success of water management projects. The national government is responsible for allocating budgets, and, accordingly, bottom-up development initiatives are often dependent on the support from the national government. Through fostering the initiation of a set of bankable projects, the program offered a solution for this impasse, especially through brokering financial deals and mechanisms for project implementation in the third phase of the program.

The challenge, however, is to ensure that the programs also receive support from financial institutions. An important step in securing funding is that, despite the lack of evaluation and funding mechanisms that are suitable for the integrated climate solutions, all programs developed in WaL Semarang must be able to present a solid business case. The early involvement of financial partners and the continuous debate about the need and usefulness of the programs helped to build the business case of the programs. The collaborative WaL process also seems to have created a sense of collective ownership of the challenges and their opportunities. In support of this, the WaL program provided the design teams with the Intellectual Property Rights as a soft guarantee for ensuring that the process will remain driven by design and that the design teams, as custodians, will also become part of the implementation.

There is still, however, the risk that only a subset of all the projects and strategic programs will receive funding and will be implemented in practice—the risk of 'cherry-picking'. This will go against the holistic character of the program since the six strategic programs are closely interrelated and are, to a certain extent, reliant upon each other to be successful. The strategic program that revolves around feeding the industry with water (Strategic Program 2 in Table 8.1), for instance, acts in concert with the program that focuses on the water-neutral industry (Strategic Program 1 in Table 8.1). For either of these projects to be successful, the implementation of the other project is needed. Therefore, the risk is that the integrated approach results in a series of separate initiatives and the funding of only a selection of the programs.

Hence, it remains to be seen to what extent the strategic programs and related climate resilient measures will be implemented. An important step in this context at the local level will be the inclusion of the programs in the next Semarang Mid-Term Development Plan, since this is compulsory for ensuring the implementation of the strategic climate resilience program for at least the next 5 years.

8.4   Conclusion

Water as Leverage has proven successful in the organization of a collaborative planning process, which has changed the mindset of the stakeholders involved. The program has demonstrated that water issues can benefit from a multidisciplinary approach and has facilitated the shaping of holistic and integrated urban climate solutions in such a way that short-term actionable interventions were integrated with long-term transformative developments. Asking the two separate multidisciplinary design teams to conduct their own analysis and design their own solutions for Semarang was pivotal since it required them to co-create these solutions. Not only has this involved a wide set of stakeholders from the start of the program, it has also led to the development of the programmatic approach. In this way, the WaL program proved that it was capable of increasing the transformative capacity of Semarang. It created an enabling environment and strengthened stakeholder capacities to stimulate the transition to a climate resilient city.

As a result, various innovative climate resilience solutions have been developed, which, when implemented, will significantly increase the city's threshold, coping and adaptive capacity. Therefore, the WaL program is an important breakthrough: it increased the awareness of local stakeholders in Semarang and at the same time activated and added a level of ambition to strengthen the coalition—and the needed structures and programs—among and across different levels of government. It further encouraged them to understand the water issues as multisectoral problems that require complex and innovative solutions and a comprehensive programmatic approach rather than a project-based approach. These lessons are also relevant to other vulnerable coastal cities that aim to collaborate, design and co-create more integrated and proactive climate solutions through a transformative process and within a 'soft and safe space' environment. An important issue that needs further investigation, however, is what will happen to these initiatives if they have to move out of these soft and safe spaces, that is, how these initiatives could be embedded in formal institutional structures.

REFERENCES

Allmendinger, P., & Haughton, G. (2009). Soft spaces, fuzzy boundaries, and metagovernance: The new spatial planning in the Thames Gateway. *Environment and Planning A, 41*(3), 617–633. https://doi.org/10.1068/a40208.

Cascading Semarang. (2018). *Cascading Semarang steps to inclusive growth phase one report.* The Hague.

De Graaf, R. E. (2009). *Innovations in urban water management to reduce the vulnerability of cities: Feasibility, case studies and governance.* PhD thesis, Technical University Delft, The Netherlands.

Dutch Special Envoy for International Affairs, Netherlands Enterprise Agency & Architecture Workroom Brussels. (2019). *Setting the scene for a call for action.*

The Hague. Retrieved from https://waterasleverage.org/file/download/57979535/waterasleverage-settingthesceneforacallforaction.pdf

Dutch Special Envoy for International Water Affairs, I. A. B. R., & Architecture Workshop Brussels. (2019). *Water as leverage for transformative impact.* The Hague.

Few, R., Morchain, D., Spear, D., Mensah, A., & Bendapudi, R. (2017). Transformation, adaptation and development: Relating concepts to practice. *Palgrave Communications, 3*(1). https://doi.org/10.1057/palcomms.2017.92.

Government of the Netherlands. (2019a). *Factsheet water as leverage.* The Hague: Ministry of Infrastructure and Water Management and Ministry of Foreign Affairs.

Government of the Netherlands. (2019b). *Team approaches.* The Hague: Ministry of Infrastructure and Water Management and Ministry of Foreign Affairs. https://waterasleverage.org/file/download/57980048/Water-as-Leverage-One-pager-team-approaches.pdf

Handayani, W., Fisher, M. R., Rudiarto, I., Setyono, J. S., & Foley, D. (2019). Operationalizing resilience: A content analysis of flood disaster planning in two coastal cities in Central Java, Indonesia. *International Journal of Disaster Risk Reduction, 35.* https://doi.org/10.1016/j.ijdrr.2019.101073.

Kaczmarek, T. (2018). Soft planning for soft spaces. Concept of Poznań metropolitan area development–a case study. *Miscellanea Geographica, 22*(4), 181–186. https://doi.org/10.2478/mgrsd-2018-0020.

Mulyana, W., Setiono, I., Selzer, A. K., Zhang, S., Dodman, D., & Schensul, D. (2013). Urbanisation, demographics and adaptation to climate change in Semarang, Indonesia. Human Settlements Group, International Institute for Environment and Development.

One Resilient Semarang. (2018). *One resilient Semarang volume 1 research, analysis and engagement (Draft report – For workshop purposes).* The Hague.

PBL Netherlands Environmental Agency. (2018). *Geography of future water challenges.* The Hague. Retrieved from https://www.pbl.nl/sites/default/files/downloads/pbl-2018-the-geography-of-future-water-challenges-2920_2.pdf

Pillai, P., Philips, B., Shyamsundar, P., Ahmed, K., & Wang, L. (2010). *Climate risks and adaptation in Asian coastal megacities.* Washington, DC. Retrieved from http://documents.worldbank.org/curated/en/866821468339644916/Climate-risks-and-adaptation-in-Asian-coastal-megacities-a-synthesis-report

United Nations, Department of Economic and Social Affairs, Population Division. (2015). *World urbanization prospects: The 2014 revision, (ST/ESA/SER.A/366).* https://population.un.org/wup/Publications/Files/WUP2014-Report.pdf

Vinke, K., Schellnhuber, H. J., Coumou, D., Geiger, T., Glanemann, N., Huber, V., Knaus, M., Kropp, J., Kriewald, S., & Laplante, B. (2017). *A region at risk: The human dimensions of climate change in Asia and the Pacific.* Manila: Asian Development Bank. https://doi.org/10.22617/TCS178839-2.

Water as Leverage (Producer). (2019). *Water as leverage team approach.* Conceptual Design presented at Semarang International Seminar.

CHAPTER 9

Making the Transition: Transformative Governance Capacities for a Resilient Rotterdam

Arnoud Molenaar, Katharina Hölscher, Derk Loorbach, and Johan Verlinde

Abstract This chapter describes how entrepreneurial policy makers, researcher and professionals in Rotterdam, the Netherlands, support the transition to a water resilient city. It identifies the capacities needed to work systematically on transformative change: stewarding, unlocking, transforming and orchestrating. This framework of transformative governance is illustrated with practical examples that show it is both possible and challenging to make the transition. A resilient city is a city that can adapt, develop and renew beyond institutional fragmentation and inertia. To work across silos, to experimentally develop and empower social

A. Molenaar • J. Verlinde
Rotterdam City Government, Rotterdam, The Netherlands

K. Hölscher • D. Loorbach (✉)
Dutch Research Institute for Transitions, Rotterdam, The Netherlands
e-mail: loorbach@drift.eur.nl

© The Author(s) 2021
R. de Graaf-van Dinther (ed.), *Climate Resilient Urban Areas,*
Palgrave Studies in Climate Resilient Societies,
https://doi.org/10.1007/978-3-030-57537-3_9

innovation and to scale and diffuse sustainable alternatives require a long-term commitment, a strategic transition perspective and a diverse set of capacities, skills and competences to come together. This chapter offers a practical and coherent perspective on how cities can work to make this transition.

Keywords Rotterdam • Transformative • Transition • Governance • Capacities • Cities

9.1   Introduction

Adapting to climate change and achieving resilience is not an easy journey. The city of Rotterdam, the Netherlands, is now working for at least 15 years on climate adaptation and has built an international profile on that topic. Rotterdam was able to become a globally leading city in climate adaptation because local politicians and civil servants recognised the urgency and the opportunities for making the city climate resilient, liveable, safe and equitable. The Rotterdam 'adaptation story' (*an impactful narrative*) does not only talk about climate change but paints a vision of a better city, a better Rotterdam. People who visit can see 'real things' (*'real applications' for dealing with climate change pressures*) such as the Benthemplein Water Square and the Floating Pavilion. Many cities that face similar struggles call civil servants and entrepreneurs in Rotterdam to learn about and from what we did.

Rotterdam is located in the north-western European Meuse river delta. Large parts of the city are below sea level, protected by intricate systems of dykes and water defence infrastructures. It has, like the Netherlands in general, a long historical relationship with water: Rotterdam was founded on high ground next to the river in the fourteenth century and has profited from as well as fought against the water since. Especially since the late nineteenth century, this relationship has been one of engineering, planning and technology-based water management. In 1872, the famous Dutch water engineer Caland helped to create the 'Nieuwe Waterweg' (new waterway) that opened up Rotterdam to the sea and made possible the development of the port. In the century to follow, masterpieces of water engineering became part of the famous Delta Works. The Maeslantkering, a storm surge barrier spanning over 400 meters that can close mechanically, created protection against flooding.

Towards the end of the twentieth century, new challenges started to emerge as climate change impacts were anticipated and increasing vulnerability of the city behind the dikes was acknowledged. Within the national discussions on water management, triggered by severe flood events in 1995 and 1996, a rethinking of water management started. The consensus shifted from 'fighting the water' to 'living with water' (Van der Brugge et al. 2005). The main reasons for this were, next to potential sea level rise, also increased (peak) rainfall and increased river flow. As construction had been allowed in flood-prone areas, there was a lack of retention space and higher likelihood of economic damages because of high water. In Rotterdam, a city and most vulnerable to all of these newly emerging threats, this spurred entrepreneurial and creative water managers and urban planners to start reimagining the city with water. During the International Architecture Biennale Rotterdam in 2005, themed 'The Flood', a new vision was presented: Rotterdam Water City 2035 (de Greef 2005).

In this chapter, we take this vision as the starting point for the transition (Loorbach et al. 2017; Hölscher et al. 2018) towards a water resilient city. It laid the intellectual and conceptual foundations for how Rotterdam could develop in a more symbiotic way with (changes in) its water system. It identified the potential of improving quality of life, creating healthy and attractive living environments at the same time while improving its adaptive capacities to deal with water stress, risks and other challenges (see Fig. 9.1). Fifteen years later, this vision is materialising in many different ways and has become part of the mainstream water management of the city. Rotterdam is widely seen as an exemplar city, also internationally, in terms of its resilience program, it innovative, small- and large-scale solutions. Its most recent policies engage the public as well as support resilient urban design. It introduces resilience thinking in relevant policy domains and actively engages in global city networks to exchange lessons and share experiences. In this chapter, we reflect upon how Rotterdam transformed its water management and the governance capacities that supported this process.

9.2 ROTTERDAM AS WATER LABORATORY

In Rotterdam, resilience became defined as a broad, cross-cutting issue that combines climate adaptation with the ambition to make Rotterdam green, liveable, equitable and healthy. This approach to climate adaptation opened many opportunities for tackling the impacts of climate change while adding value. At the same time, it created a basis for interaction and

Fig. 9.1 Rotterdam, a global leader in climate adaptation located in a low-lying river delta (Gerhard van Roon/ Kunst en Vliegwerk)

sometimes confrontation with other policy domains more focused on economic and spatial development for example. A lot of the integrated and transformative resilience vision can be traced back to the central ideas of Rotterdam Water City 2035 presented in 2005—viewing water as opportunity to add value to the city and testing new multi-functional infrastructures for water retention. By now Rotterdam has over 400,000 m² of green roof space that also improves the air quality of the city and mitigates flood risk. Its 'water squares' that serve as retention areas as well as add spatial quality are copied across the world (see Fig. 9.2). Innovations like these supported new business models and local companies are now involved in the development of water squares in Surat, India. The new program 'Rotterdams Weerwoord' (Gemeente Rotterdam 2019) engages citizens in public debates and small-scale initiatives like greening gardens

Fig. 9.2 Creating multi-functional resilient public space. (Picture: Guido Pijper)

and harvesting rainwater. Resilience strategies have become part of infrastructure development, housing policies and nature development.

Water issues have thus become closely linked to climate adaptation and other policy priorities. This has been an explicit goal of Rotterdam's water strategy from the beginning: to broaden the 'water management sector' to bring more water sensitive and water resilient thinking into other policy domains related to spatial quality, urban development, infrastructure and spatial planning. In 2007, the Rotterdam city government launched the Rotterdam Climate Initiative (RCI) focusing on CO_2 reduction through mitigation. In 2008, the Rotterdam Climate Proof (RCP) programme was added as a response to calls from the water management community to bring in adaptation as part of the RCI. It included the goal to make Rotterdam 100% climate proof against flooding and the urban heat island effect by 2025. This RCP took up the reframing of urban water management and climate change as contributing to the quality of the social environment in the city: while becoming climate resilient, additional opportunities to enhance the attractiveness of the city in terms of living, recreation, working and investments should be realised. The 2012 published Rotterdam Adaptation Strategy (RCI 2012) serves as framework to develop collaborative climate adaptation measures.

In the decade to follow, the program developed all sorts of initiatives, projects an, activities and communication around the idea of transition to a water resilient city. Characteristic to this 'Rotterdam approach' is that transformative actions and projects have been developed and implemented in parallel to developing transformative networks and strategies—long-term visions can this way be linked directly to immediate examples on the ground which in turn make the integrated approach visible and tangible. Examples of multi-functional innovative solutions include the Benthemplein Water Square, which combines rainwater management with area development, the multi-functional underground water storage facility at Museumplein car park and the Floating Pavilion. The Dakakkers is the first multi-functional rooftop garden in Rotterdam, combining flood protection with commercial and recreational use (see Fig. 9.3). The levee at

Fig. 9.3 Benthemplein Water Square functions as social space and retention area. (Picture: Arnoud Molenaar)

Vierhaven area was expanded to include a city park, and a number of new 'tidal park' areas are planned along the rivers in the Rotterdam region. The strategy is now to combine measures on a small scale, promoting in particular small projects by citizens and businesses, with few 'eye-catchers' and effective large-scale projects operating in the background.

With these practical initiatives and high-profile proof-of-concept experiments, Rotterdam has gained international recognition. Increasingly, international delegations visit Rotterdam and scientific conferences are held in the city. Due to the international profiling, the city has struck up international partnerships for knowledge exchange, and businesses from the Rotterdam region have become active in cities like Ho Chi Minh City, Jakarta and New Orleans. The participation in the Rockefeller Foundation's 100 Resilient Cities programme helped to create such international attention for Rotterdam. The strategic approach for resilience was then further institutionalised in the city government's cross-cutting Sustainability department, the Water department and Resilience team and the appointment of a Chief Resilience Officer in 2014, and Rotterdam becoming the hosting city for the UN Global Centre on Climate Adaptation in 2020.

Over the past two years, the Resilience Strategy broadened Rotterdam's focus from climate adaptation to also include social resilience, the resilience of vital infrastructure and cyber security (Gemeente Rotterdam 2016). It also helped diffuse the attention for water management and resilience beyond the professional water management sector making clear that local communities could be severely affected by water risks. Especially vulnerable communities would be exposed to such risks without the means or preparation to deal with them. This led to the development of social resilience strategies focusing on creating awareness, local adaptation projects and capacity building. Resilience also opened up to broaden the scope from water to other interrelated risks: a number of risks started to emerge related to the digitalisation of water management infrastructures as well as the economy in broader sense. This led to the development of a strategy for cyber resilience, applying resilience thinking to digital infrastructures and risks. For example, the strategy identifies critical roads and infrastructures to be protected, links water safety to cyber security and identifies community initiatives that could be connected to the city's resilience efforts.

9.3   Developing Transformative Governance

The shift from more traditional water management towards the resilience strategy cutting across policy domains and engaging policy, market and public was supported by a broad range of strategic governance activities supporting learning, experimentation, visioning and coalition building. It was learned that governance could not only involve more traditional policy instruments like planning, subsidies and regulations, but also require changes in discourse, behaviour and wider institutional settings. This experience highlights the need to not only think about new solutions and innovations to create water resilient cities, but that we must also think about new forms of governance to support their implementation and diffusion. In other words: if water management is a separate policy domain not addressing the production of fragmented and *unresilient* spatial policies, it will not succeed in supporting a shift towards resilient cities. Hölscher (2019) identified, based on empirical work in New York City and Rotterdam, combined with scientific work on adaptive and transformative governance, the governance capacities needed (Fig. 9.4). We here reiterate the framework and summarise its core messages.

The starting point for identifying transformative governance capacities is that achieving a resilient city is not a straightforward process that can be planned or implemented: it will require continuous, collaborative and coordinated processes of experimentation, innovation, change and learning. It will be different in every city, district or location, but the resilience principles and thinking in itself can act as a more generic and universal or 'translocal' framework (Loorbach et al. 2020). Resilience is understood as not only about climate adaptation and climate proofing, but about digging into the deeper drivers of vulnerabilities and risks such as land use planning, community life and lifestyles, and to advance ways of thinking that put principles as diversity, flexibility and adaptive capacity front and central. Given the more fragmented, institutionalised and efficiency dominated modes of urban planning, resilience ultimately is about systemic change in existing societal structures, institutions and regimes to improve, maintain and protect social and environmental wellbeing in the long-term.

The Resilience Strategy in Rotterdam has to a large extent achieved to open up existing more domain-specific and 'rigid' societal structures, creating more space and impact of concerted decision-making across sectors and scales, political leadership and collaborative and learning-based decision-making. In hindsight, four capacities helped to create this impact

Fig. 9.4 Dakakkers: an urban garden on the roof. (Photo: Ossip van Duivenbode)

and facilitate transformative change towards resilient cities. These capacities for transforming governance are oriented towards long-term goals, target root causes rather than symptoms, foster inclusive collaboration across sectors and scales, and allow for continuous learning to innovate, adapt and scale (Hölscher 2019; Fig. 9.5). Rather than controlling change, these transformative governance capacities facilitate responses that mobilise and respond to the dynamics influencing change, inertia and risks in cities. Governance capacities continuously evolve from governance agents creating, mobilising and changing the institutional, organisational, social and knowledge conditions influencing how and what kind of decisions they can make.

These four capacities each play a role in developing, diffusing and anchoring a new discourse, structures, relations and practices in a context

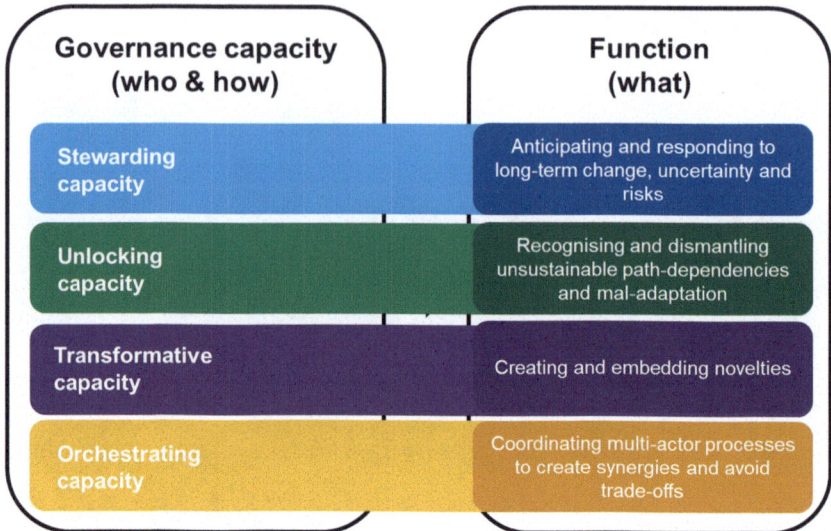

Fig. 9.5 Transformative governance capacities for resilient cities. (Adapted from: Hölscher 2019)

where an institutionalised discourse and practices are argued to no longer work on the longer term. From a sustainability transitions perspective (Loorbach et al. 2017), this is referred to as a persistent problem in a societal regime: a historically evolved dominant culture, structure and practice in a specific societal system or area. In this case: the historic Dutch water management regime that evolved as a separate policy domain dominated by engineers and planners. As this 'regime' was increasingly found to be unsustainable on the longer term, it implied a simultaneous transformative challenge for water management itself (to open up and develop more adaptive, ecological, resilient strategies) as well as for other policy domains to pay attention to water resilience in their policy development. These four capacities identify the different types of impact governance can have to support such a broader societal transition, and we illustrate this with how they were developed and became manifest in Rotterdam.

9.3.1 Stewarding Capacity in Rotterdam

Stewarding capacity facilitates flexible responses to deal with uncertainty and protect and recover from risks and surprises such as heavy storms or heat waves. Rotterdam has invested in stewarding capacity by creating a vast knowledge base about (future) water and climate-related risks and vulnerabilities. National, regional and international knowledge programmes and partnerships support knowledge generation. For example, Knowledge for Climate, a Dutch research collaboration, and the public–private National Delta Programme contributed to research on climate risks and adaptation strategies (van den Berg et al. 2013; van Veelen 2013).

Partnerships and collaborations are an important condition underpinning stewarding capacity. Public–private partnerships such as the RCP or neighbourhood-based planning processes promote collaboration between public and private partners for the development and implementation of concrete projects. Pilot studies for climate proofing are developed together with local communities, which help to clarify problems and constraints of the local government—such as national regulation or lack of financial means—as well as to address the needs of local communities.

9.3.2 Unlocking Capacity in Rotterdam

Unlocking capacity removes the root causes, including existing infrastructures, values, behaviours and power relations, that drive high emissions and unsustainability. Unlocking capacity in Rotterdam is visible in the ability to create a new narrative about the city's approach to managing water, particularly in moving away from perceiving water as a threat and the aim to create space for living with water. The development of the floating pavilion in the Rijnhaven, a water body directly connected to the sea, contributes to this narrative, showcasing that floating buildings are part of the answer to climate change for coastal cities.

The generation of knowledge about drivers of risk and vulnerabilities helps to identify target areas for action (see e.g. Ligtvoet et al. 2015; RCI 2012). It served to trigger urgency, broaden the perspectives of decision-makers and facilitate robust decision-making and investments. The knowledge generated underscored that the impacts of climate change, including sea level rise but also changing rainfall patterns, will make it impossible to continue relying on existing drainage infrastructure. Institutional changes such as changes in planning regulation, operational standards and incentive structures are an important condition to make new practices more beneficial. Examples are the provision of incentives to property owners to invest in green roofs (Mees et al. 2013).

Political and societal support has been critical for increasing opportunities to depart from business-as-usual approaches. The increasing awareness about the impacts of climate change (and the discourse on water as opportunity) has led to stronger public and private willingness to invest in climate adaptation, and supports cooperative spatial planning and flood management. Key activities to motivate residents and businesses to take action on their own plot of land are online awareness campaigns, information evenings and door-to-door information leaflets that give practical examples of how to make houses and plots more water proof. However, it is still challenging to engage land owners and residents in changing their social practices to pave their gardens. While residents are responsible for rainwater on their own plots, residents are often still unaware of their risks and responsibilities and rely on the government to be protected against flooding risks. A challenge in this approach also lies in the need for programmatic continuity to become a water and climate resilient city. In addition to stewarding capacity, the growing (inter)national profile helps to keep the topic and programmatic approach on the political agenda.

9.3.3   Transformative Capacity

Transformative capacity enables the development and spreading of new narratives, solutions and practices that provide alternatives for creating sustainable and resilient cities. Policy entrepreneurs in Rotterdam were able to create informal and protective spaces, in which relatively small groups of public and private actors from different governance levels could come together to share knowledge and develop innovations. Because discussions were framed strategically as 'non-official', flexible and informal, more radical ideas and longer timeframes were possible than in usual policy making practice. In the mid-2000s, policy entrepreneurs used international momentum to reframe the city's water management approach from 'keeping water out' towards 'water as opportunity for liveability'. In this context, the International Architecture Biennial in 2005, with the theme 'The Flood', certainly was such a momentum and the 'Rotterdam Water City 2035' vision created during that event became a driver of change.

These open-ended innovation collaborations were institutionalised in new partnerships and networks for ongoing experimentation. The momentum was sustained through the Rotterdam Climate Proof program in 2008, addressing and implementing adaptation in a programmatic way. And it was given a new impulse in 2014 with the membership of the 100 Resilient Cities network, supported by the Rockefeller Foundation. These larger programmes helped to mainstream and diffuse new ideas, practices and transform institutional structures. In addition, lessons learned from implementing proof-of-concept experimental projects support their replication and upscaling, ranging from green roof subsidy schemes and construction regulations to ecological green infrastructures. For example, the lessons learned from the complex maintenance of the Benthemplein water square were translated in improved and more standardised designs. However, so far the innovative goals and solutions remain largely disconnected from conventional planning and decision-making processes, pointing to limited mainstreaming. Learning from practical experiments remains informal due to time constraints—no explicit attention is given to reflexivity and monitoring.

9.3.4   Orchestrating Capacity

Orchestrating capacity is about the coordination of actors to ensure that all their activities are aligned towards shared long-term goals and in this way create stepping stones and synergies for transformations. The long-term sustainability and resilience goals guide governance activities in the city and filter through into other policy domains. For example, the programme 'River as Tidal Park' to strengthen the Meuse river as central, green space connects economic activity, greening, biodiversity and recreation and is implemented by the port authority, the city government and environmental organisations.

To coordinate the implementation of the strategic agenda, the Rotterdam Water and Adaptation department, the Sustainability department and the Resilience team are all tasked with motivating, overseeing and coordinating planning processes across sectors. Their cross-departmental set-up makes them central nodes for knowledge exchange and pooling. The offices' policy officers initiate and organise joint visioning processes, identify opportunities for experimentation and piggybacking climate mitigation and adaptation initiatives, search and allocate funding sources and participate in cross-scale collaborations and international city networks. The position of the Chief Resilience Officer provides a key contact point for pooling all resilience efforts in the city.

9.4   Conclusions and Outlook

As Rotterdam is moving into the new decade, the key challenge that remains is the further mainstreaming of the city's approach to climate adaptation and resilience. With the start of the UN Global Centre of Excellence on Climate Adaptation and the Climate Adaptation Academy in 2020, the city gets another momentum for its transition. After building up the governance capacities at least to a basic level, the resilience discourse and a portfolio of solutions, technologies, models and approaches, it now is entering the phase of mainstreaming and structural change. The big question is the extent to which the resilience governance network can sustain the momentum in times of crises and economic pressures. The more informal, distributed and entrepreneurial, experimental way of working might have been successful getting this far, but the question is whether it is enough to really make the transition.

Within the city government, those working on resilience operate across the different policy departments rather than that they are embedded within the institutional structures. The visions and programmes do not yet pervade to the operational level of city departments' everyday work. It still has not become business as usual to develop the city based on resilience principles. Most power still lies in the budgetary systems that give limited opportunities for long-term and synergistic financing. This will need to be considered in the update of the resilience strategy from 2016: the goals are clear, but the next steps need attention to *how* to achieve them in terms of institutionalising resilience and climate adaptation, developing wider cross-cutting partnerships, connecting with community groups and on-going reflexive monitoring to learn about what works and what does not to achieve resilience in the long-term.

Looking beyond Rotterdam, it is inspiring to see other cities around the world work on similar challenges. Cities are growing together as a community to join forces, exchange ideas and learn from one another. More and more cities experience the value of peer-to-peer exchange, which is one of the benefits of being a member of city networks like the C40, Covenant of Mayors and 100 Resilient Cities. These are great networks to exchange and inspire transformative governance capacities to achieve full transition. For Rotterdam, this implies a broadening of our scope and an increase in our institutional impact. Our scope needs to widen to include economic, cyber and social resilience as intrinsically linked to a resilient city. To address institutional fragmentation and the potential value in working based on resilience and synergy in local area development. To mobilise the public to create support and local engagement in resilience measures. To develop training and education for professionals to help develop and diffuse transformative governance capacities across policy domains and between policy and practice. And to invest more deeply in the scaling and institutionalisation of governance capacities needed for the sustainability transition. Achieving a resilient city by itself might always be a moving target, but making the transition in a structured, collaborative and learning-by-doing way will bring us closer and closer.

References

De Greef, P. (Ed.). (2005). *Rotterdam Waterstad 2035*. Rotterdam: Jap Sam Books.
Gemeente Rotterdam. (2016). *Rotterdam resilience strategy. Ready for the 21st century*. http://lghttp.60358.nexcesscdn.net/8046264/images/page/-/100rc/pdfs/strategy-resilient-rotterdam.pdf. Accessed 20 Sept 2016.
Gemeente Rotterdam. (2019). *Rotterdams WeerWoord*. Urgentiedocument. https://www.rotterdam.nl/wonen-leven/rotterdams-weerwoord/Urgentiedocument-2020_NL.pdf
Hölscher, K. 2019. *Transforming urban climate governance: Capacities for transformative climate governance*.
Hölscher, K., Wittmayer, J. M., & Loorbach, D. A. (2018). Transition versus transformation: What's the difference? *Environmental Innovation and Societal Transitions, 27*, 1–3.
Ligtvoet, W., van Oostenbrugge, R. J., Knoop, J., Muilwijk, H., & Vonk, M. (2015). *Adaptation to climate change in the Netherlands – Studying related risks and opportunities*. The Hague: PBL Netherlands Environmental Assessment Agency. http://www.pbl.nl/sites/default/files/cms/publicaties/PBL-2015-Adaptation-to-climage-change-1632.pdf
Loorbach, D., Frantzeskaki, N., & Avelino, F. (2017). Sustainability transitions research: Transforming science and practice for societal change. *Annual Review of Environment and Resources, 42*(1).
Loorbach, D., Wittmayer, J., Avelino, F., Von Wirth, T., & Frantzeskaki, N. (2020). Transformative innovation and translocal diffusion.
Mees, Heleen L.P., et al. (2013). Who governs climate adaptation? Getting green roofs for stormwater retention off the ground. *Journal of Environmental Planning and Management, 56*(6), 802–825.

RCI. (2012). *Rotterdam climate change adaptation strategy*. http://
www.rotterdamclimateinitiative.nl/documents/2015-en-ouder/
Documenten/20121210_RAS_EN_lr_versie_4.pdf. Accessed 10 June 2016.

Van den Berg, H., van Buuren, A., Duijn, M., van der Lee, D., Tromp, E., & van
Veelen, P. (2013). *Governance van lokale adaptatiestrategieen, de casus
Feijenoord*. Kennis voor klimaat. KvK report 103/2013.

Van der Brugge, R., Rotmans, J., & Loorbach, D. (2005). The transition in Dutch
water management. *Regional Environmental Change, 5*(1), 164.

Van Veelen, P. (2013). *Adaptive strategies for the unembanked area in Rotterdam*.
Synthesis report. KvK report HSRR3.1 2013.

CHAPTER 10

Future Outlook: Emerging Trends and Key Ingredients for the Transition to Resilient Delta Cities

Rutger de Graaf-van Dinther

Abstract Five capacities of urban climate resilience are discussed. These capacities are threshold capacity, coping capacity, recovery capacity, adaptive capacity and transformative capacity. Key processes for the transition to resilient delta cities include: implementing innovations at appropriate scale and speed; transdisciplinary planning and collaboration; capacity building at a local level; and to move from sustainability to regeneration. At a fundamental level, building transformative capacity is about changing water management, urban development and urban redevelopment practice. The main objective is to provide different outcomes than conventional practice. This requires transformation of society itself. A potential way to build knowledge and experience in this field is by creating urban

R. de Graaf-van Dinther (✉)
Research Centre Sustainable Port Cities, Rotterdam University of Applied Sciences, Rotterdam, The Netherlands

Blue21 and Indymo, Delft, The Netherlands
e-mail: r.e.de.graaf@hr.nl

projects and processes where new societal models based on different values and assumptions could be tested to provide more resilient outcomes. Floating communities can provide a testing space for this.

Keywords Climate resilience • Transformation • Transformative capacity • Floating communities

The idea for this book emerged in 2019 when the editor of this book wrote a short opinion article (De Graaf-van Dinther 2019) about the four capacities to increase the climate resilience of urban areas. This article was based on his PhD thesis (De Graaf 2009). The four capacities are threshold capacity, coping capacity, recovery capacity and adaptive capacity. Following this publication, Henk Ovink (2019), the Special Envoy for International Water Affairs at Kingdom of the Netherlands, suggested to add a fifth component: transformative capacity. In this book (De Graaf-van Dinther 2020), the five capacities framework of climate resilience was explored by inviting authors to write and reflect on one or more of the capacities based on their research and experience. Table 10.1 shows how the capacities of climate resilience were addressed by the contributing authors. Some authors addressed all capacities, whereas others focused on a few capacities.

10.1   Reflection on the Five Capacities of Climate Resilience

An interesting question is how the five capacities are related to each other. One way is to interpret the five capacities as equally important and similar components of climate resilience. From this perspective (Fig. 10.1a), five capacities should be strengthened in order to achieve climate resilience. In urban climate resilience strategies, all capacities should be taken into account, and none of the capacities should be neglected. In this way, the framework could also be used to evaluate urban resilience strategies and assess if all capacities are sufficiently represented. A second possible viewpoint is to consider adaptive capacity to include threshold capacity, coping capacity and recovery capacity at different spatial scales and temporal scales. Adaptive approaches could for instance include a wide

Table 10.1 Overview of how the five capacities of climate resilient were addressed in the chapters of this book

Chapter	Authors	Threshold capacity	Coping capacity	Recovery capacity	Adaptive capacity	Transformative capacity
1	De Graaf-van Dinther & Ovink	✓	✓	✓	✓	✓
2	Dolman	✓	✓	✓	✓	✓
3	Kluck & Boogaard	✓	✓		✓	
4	Escarameia & Tagg		✓	✓		
5	Van de Ven, Hooijmeijer and Storm			✓		✓
6	Rijke, Geerling, Quan & Trung			✓	✓	✓
7	French, Trundle, Korte & Camari	✓	✓	✓	✓	✓
8	Laeni, Ovink, Busscher, Handayani & Van den Brink	✓	✓		✓	✓
9	Molenaar, Hölscher, Loorbach & Verlinde				✓	✓

variety of threshold capacity and coping capacity increasing measures at building level and street level. At an urban level this would translate in a more flexible, decentralised and more adaptable urban system. Transformative capacity would then include the other four capacities but additionally address the root causes of climate change, include ecological harmony and empower local communities. Figure 10.1b illustrates this second viewpoint. The third understanding of the five capacities would be to see it as a transition (Fig. 10.1c). The assumption is then that cities generally start with implementing threshold capacity measures, then start working on coping and recovery capacity. When this is insufficient to deal

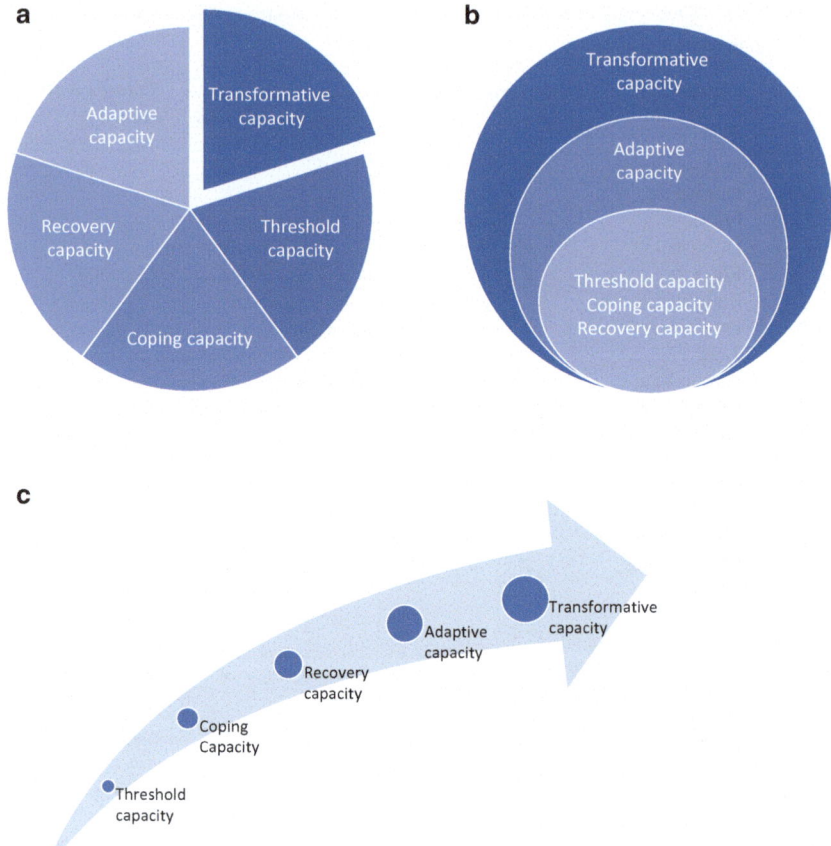

Fig. 10.1 (a–c) Possible relations between the five capacities of urban climate resilience

with the challenges they are facing, they start developing adaptive strategies until finally transformative capacities are developed. The threshold capacity phase is characterised by an environmental control paradigm which can be labelled *Fighting against Water*. The coping and recovery phase is about *Living with Water*, whereas adaptive capacity is characterised by *Anticipating on Water*. The Transformative capacity phase can be called *Water Sensitive Society*. When we take into account the severity and urgency of urban climate impacts, this raises the important question whether it is possible to leapfrog certain phases in this transition and start developing transformative capacity when threshold, coping and recovery capacities are not yet developed.

This book does not provide a conclusive answer to which of the three viewpoints is most accurate. However, exploring the five components of climate resilience and their possible relations provides a more comprehensive understanding of the holistic nature of the concept of resilience. This book also provided more insights about the suggested fifth capacity of the resilience framework. Most contributing authors acknowledge that the fifth capacity, transformative capacity, is needed to deal with the global urban climate crisis. With the results of the different chapters, it is possible to extract some key ingredients for the transition towards transformative climate resilience.

10.2   Overview of Key Transition Ingredients

One of the key ingredients in climate resilience is integration of water management and urban planning. Dolman gave an overview in Chap. 2 about various design and planning tools that are used to move from a reactive approach towards regenerative urban ecosystems. When addressing urban water challenges, measures need to be linked to ecological services, for instance, by providing blue-green infrastructures (such as green roofs, bioswales, porous pavement and water squares). The role of involved and aware communities and collaborative spatial planning is crucial besides technical measures to achieve the attractive goal of liveable and attractive green-blue cities.

For resilient spatial planning approaches to be effective in contributing to transformation at an urban scale, they need to be implemented at street level in existing cities. In Chap. 3, Kluck and Boogaard consider climate adaptation as part of street reconstruction and urban retrofitting projects. The results show that if the total life cycle costs are taken into account, climate resilient measures are not necessarily more expensive than conventional measures. Awareness of alternative solutions among professionals and citizens is crucial for these kinds of projects to be effective. Showing real and inspiring examples and the use of web-based knowledge sharing platforms can contribute to this purpose.

Critical urban infrastructures are assets that are essential for the continuity of economic activities and for meeting the fundamental needs of the urban population during flood events and other disasters. In Chap. 4, Escarameia and Tagg report on different methods for flood damage estimation, resilience assessment of critical buildings and the effectiveness of resilience-enhancing methods. These methods can in particular be effective to improve the coping and recovery capacities of urban centres when impacted by floods.

Reconstruction after disasters offers the chance to increase resilience by "Building Back Better". In Chap. 5, Van de Ven, Hooijmeijer and Storm demonstrate in their study of the recovery processes in Japan from the Tohoku Tsunami in 2011 and on the Bahamas from hurricane Dorian in 2019 that this can only be done by organising a comprehensive, multi-disciplinary planning and redevelopment process in which both experts and local stakeholders are involved. The main challenge is to build a collaborative design process with a vision for the social recovery of the community in circumstances under heavy time pressure to reconstruct the physical infrastructure and housing as quickly as possible.

The concepts of resilience and the circular economy could strengthen each other. Rijke et al. explore this in Chap. 6 for the city of Can Tho, Vietnam, where plans were developed to recycle plastic waste to construct climate adaptive floating islands for agriculture and aquaculture. In the design process, the resilience concept provided the scientific background and circular economy concept provided the more practical implementation perspective, creating livelihood opportunities for vulnerable urban communities.

The climate urban resilience crisis is predominantly pressing and urgent for the one billion people living in informal settlements in the Global South. French et al. demonstrate in Chap. 7 that by operationalising resilience through a climate justice lens, informal settlements could be upgraded and transformed as part of efforts to generate climate resilient urban areas in coastal cities. For transformative capacity, the chapter identified five "ingredients": socio-technical innovation; a climate justice framing; greater attention to intersectional dimensions; inclusive governance and community empowerment; and fit for purpose finance. Transformative capacity can offer a pathway for informal settlements to foster more significant material improvements in human health and conditions. This reflects a broader need for climate change responses to reform the prevailing urban development models and approaches, which perpetuate structural inequalities and unsustainable urban development practices.

A transformative process for urban climate resilience in Semarang, Indonesia, as part of the Water as Leverage programme is described in Chap. 8 by Laeni et al. A key ingredient was building local system knowledge and joint system understanding by an inclusive development process of multi-disciplinary expert groups and local stakeholders. Innovative and unexpected multi-sector solutions were developed in a holistic and integrated programme. Another success factor was creating an enabling environment and strengthening stakeholder capacities by facilitating formal and informal collaboration at different levels of scale. The programme aims at securing sustainable financing mechanisms by early involvement of financial partners and developing solid business cases.

Molenaar et al. discuss the experience of the city of Rotterdam, the Netherlands, with transformative governance. Rotterdam has become a "water laboratory" for a wide range of innovative resilience projects. This has strengthened the local awareness for water management as well as the international recognition of Rotterdam as a frontrunner in climate resilience. A key success factor has been to link water and climate issues to other policy priorities such as spatial quality. The next step for transformative governance is to broaden the scope from climate resilience to include other domains such as social resilience. Institutional fragmentation should be addressed, and local engagement and mobilising the public are essential ingredients for a successful transition. Professional training and capacity building are needed to develop and diffuse transformative governance capacities across policy domains and between policy and practice. Investments for scaling and institutionalisation of transformative capacity is needed.

10.3   The Transition Outline: From Adaptation to Transformation

The different chapters in this book demonstrate that we are in a transition in how urban areas respond to climate threats and internal social dynamics. A number of required processes can be observed by analysing common elements in the chapters (Table 10.2). The size and urgency of climate impact is such that adaptation is no longer considered sufficient. Instead, transformation of the entire urban system is needed to anticipate on climate change impacts.

Table 10.2 Key required processes for the transition from adaptation to transformation

	De Graaf and Ovink	Dolman	Kluck and Boogaard	Escarameia and Tagg	Van de Ven et al.	Rijke et al.	French et al.	Laeni, Ovink et al	Molenaar et al.
Implementing innovations at appropriate scale and speed	1	2	3	4	5	6	7	8	9
Blue-green infrastructures	✓	✓	✓						✓
Floodproof critical infrastructure	✓			✓					✓
Floating developments	✓				✓	✓		✓	✓
Upscaling and replication	✓		✓					✓	✓
International knowledge sharing of best practices	✓	✓	✓			✓		✓	✓
Total life cycle costs and business cases	✓		✓	✓	✓	✓		✓	
Project based to process based	✓	✓	✓				✓	✓	✓
Planning and collaboration	1	2	3	4	5	6	7	8	9
Informal policy processes	✓	✓			✓	✓		✓	✓
Transdisciplinary collaboration	✓	✓			✓	✓	✓	✓	✓
Coalitions of experts and citizens	✓	✓	✓		✓	✓	✓	✓	✓

Capacity building at a local level	1	2	3	4	5	6	7	8	9
Community empowerment	✓	✓			✓	✓	✓	✓	✓
Community-based financing							✓	✓	
Provide livelihood by resilience initiatives						✓	✓		
From sustainability to regeneration	1	2	3	4	5	6	7	8	9
Restore damaged ecosystems	✓				✓	✓	✓	✓	✓
Urban development in harmony with nature	✓	✓				✓	✓	✓	✓
Provide ecological services	✓	✓				✓	✓	✓	

10.3.1   Implementing Innovations at Appropriate Scale and Speed

The transition from adaptation to transformation requires innovative new technical measures, such as blue-green infrastructures (e.g. green roofs, bioswales, porous pavement and water squares), floodproofing critical infrastructures and floating developments. More flexible and adaptable urban infrastructures are needed to be able to prevent a technical and institutional lock-in of centralised capital-intensive urban infrastructures that is not sufficiently equipped to adapt to rapid social and environmental change. For innovations to have a system-wide impact on an urban scale and global scale, they need to be applied at an appropriate scale and speed which is relevant and not symbolic compared to the magnitude and severity of urban climate change impacts and urban growth processes to which they aim to respond. Upscaling, replication and mainstreaming of innovations are key processes that are needed for this purpose. Global sharing of best practices by partnerships and online knowledge platforms are tools that can contribute to increasing the scale and speed of climate resilient transformation. Monitoring, learning and evaluation of innovative climate projects are needed to enable innovations to move from a niche-scale project-based approach towards an inclusive process-based approach in which innovations become able and equipped to compete with large-scale centralised unsustainable infrastructures. Ideally, transformative climate resilience would become an integral part of every intervention in the urban environment. Assessment of total life cycle costs, system-wide impacts and developing solid business cases could assist in arranging the much larger required budget to implement transformative climate resilience at an appropriate speed and scale. Currently, innovations are still mainly implemented as small-scale innovative niche projects, while at the same time, the status quo is reinforced at a much larger scale.

10.3.2   Planning and Collaboration

Social and ecological considerations become increasingly important compared to the conventional, mainly technical, planning approach. This is one of the essential elements of transformative capacity. The required transition cannot be forced by top-down hierarchical planning. Inclusive transdisciplinary planning with local communities and experts is needed to mobilise the required local knowledge and support for the climate resilient transformation process. The role of citizens changes from a passive consumer towards a co-producer and a source of context-specific local knowledge. The role of experts is to provide reliable knowledge and work together with other disciplines to provide integrated knowledge beyond their own expertise to enable system innovation and system transformation. Informal policy processes can offer a mechanism and a protected space for stakeholders to come up with innovative solutions that would not be accepted in the context of conventional mainstream formal policy processes. By creating momentum, societal acceptance and political relevance, these informal processes can start to influence the mainstream policy cycles and create an enabling mechanism for mainstreaming transformative innovations.

10.3.3 Capacity Building at a Local Level

Transformative climate resilience is characterised by community-based local urban systems. Citizen empowerment and capacity building among local stakeholders are crucial success factors for the transition to climate resilient urban areas. Citizens should be empowered to engage in design and decision-making processes which shape their living environment. This also includes improving access to, and increasing the size of, community-based financing as an opposite to the currently most dominant top-down financing mechanisms which reinforce vested interests and the status quo. The governance capacity of local stakeholders needs to be strengthened to enable them to design, implement and maintain the developed technical measures. Linking climate resilience to initiatives that provide livelihood to local communities can be a promising strategy to involve local communities in climate resilience initiatives.

10.3.4 From Sustainability to Regeneration

Part of the required transition to create climate resilient urban areas is the move from sustainability to regeneration. Urban sustainability policy is still predominantly characterised by reducing, mitigating and compensating the negative ecological impact of urban areas on the environment. With climate change impacts and urban growth, this approach will not be sufficient to preserve functioning ecosystems in the future which both have a significant intrinsic value as well an essential function in providing ecosystem services to urban areas. Urban development and urban redevelopment should provide housing, infrastructures and public space in harmony with nature. The word regenerative implies that urban development and redevelopment processes should also contribute to restore damaged ecosystems, by providing habitat and linkages to other nature areas. Additionally, urban development should address the root cause of climate change by creating urban areas that have a net negative impact on CO_2 emissions by capturing greenhouse gasses from the atmosphere.

10.4   Transformation Towards a New Societal Model

At a fundamental level, building transformative capacity is about changing water management, urban development and urban redevelopment practice. The main objective is to provide different outcomes than conventional practice. Sustainable urban transformations can be characterised by sustainable places, the sustainable transition of the urban development regime and the sustainable transition of related societal sectors such as energy and transportation (Ernst et al. 2016). Current models of urban development are still based on nineteenth and twentieth century values, even if they are applied in an innovative way. Examples of dominant values are: public health and safety, centralised environmental control, individualism, cost efficiency, expertise specialisation and separation, and citizens as passive consumers (consumerism). Innovative urban development at the required and relevant scale to deal with the current urban climate crisis is not possible without addressing these dominant and underlying assumptions and values. As a consequence, one should conclude that not a transformation of urban development and urban redevelopment is needed, but a transformation of society itself. A potential way to build knowledge and experience in this field is by creating urban projects and processes where a new societal model—based on different values and assumptions—could be tested such as Urban Living Labs (Von Wirth et al. 2019). Relevant values and assumptions for such developments could include community empowerment, environmental justice, social equity, transdisciplinary design by experts and citizens, integrated system design, regenerative urban developments and local community-controlled sources of water and energy.

10.5   FLOATING COMMUNITIES AS TESTING SPACE FOR TRANSFORMATIVE RESILIENCE

The idea of creating floating cities as a potential response to the climate change impacts on coastal cities has been gaining momentum in the past 10 to 15 years. Last year, UN Deputy Secretary-General Amina Mohammed stated that sustainable floating cities can offer solutions to climate change threats facing urban areas (UN 2019). The obvious reasons that are often mentioned are sea level rise and land subsidence in many coastal cities, while global land scarcity might be even more urgent (Roeffen et al. 2013). However, floating urban areas can also provide a promising context for scientific research (Marris 2017), including experimenting with new societal models (De Graaf 2012; Quirk 2017). For transformative climate resilience to be strengthened, these experiments should be based on different values, for instance the ones that were outlined at the end of the last paragraph. Floating communities could then provide a learning environment for leap-frogging towards *Water Sensitive Societies* by strengthening transformative capacity. In these floating communities, affordable floodproof floating houses should be provided for involved and climate aware communities. The design should be based on the local cultural context (Trang 2016). Citizens' empowerment could be increased because these communities are not connected to centralised infrastructure controlled by large utility companies. Citizen can organise their own local scale water and energy supply based on decentralised technologies. Citizens can also be involved in designing their own living environment assisted by experts who provide the most recent knowledge. Floating communities can have a negative CO_2 footprint if they are surrounded by floating aquatic biomass and will be able to provide food security, for instance by aquaculture (Dal Bo Zanon et al. 2017). Energy and ecosystem services can be provided to neighbouring coastal cities, for instance by treating wastewater and CO_2 with floating technologies (Trent et al. 2010). To create such a symbiotic relation between the floating community and the coastal city, the focus should not be to create floating buildings, floating houses and floating platforms as this is mostly established technology (Wang et al. 2020). Instead, the main focus should be the development of floating green-blue infrastructure (Fig. 10.2). This will enable the development of water-based ecosystem services and the provision of these services to coastal cities. The cultivation of floating aquatic biomass and food production will also provide livelihood

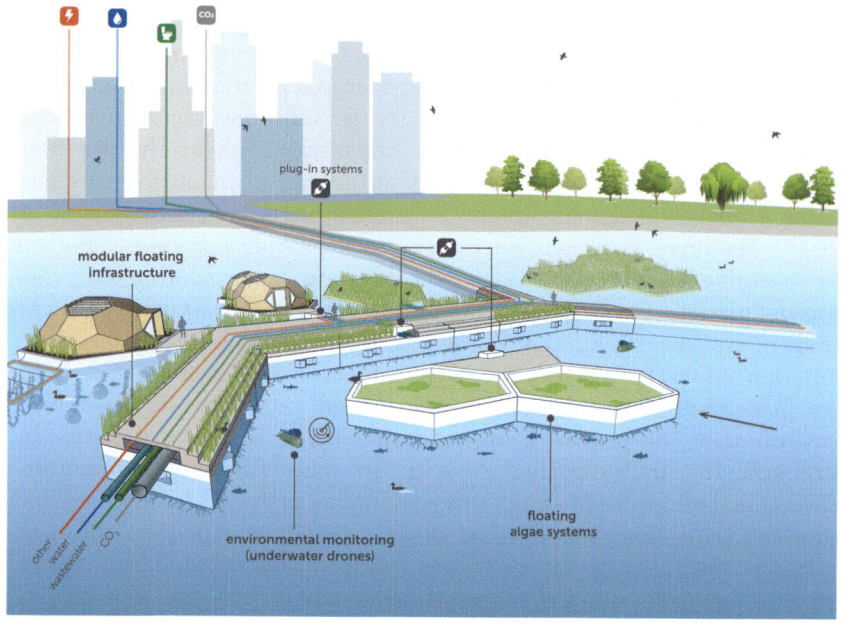

Fig. 10.2   Floating infrastructure enables the creation of floating communities as testing space for transformative resilience (Design: Blue21)

opportunities for local communities. Under floating structures new aquatic ecosystems can emerge (De Lima et al. 2015). The impact of floating communities on the environment can be monitored real-time with aquatic drones (De Lima et al. 2020). The built experience with innovative local technologies, new social constellations and harmonic relations between humans and nature can be translated back to land-based communities in coastal cities, in order to accelerate the transition to transformative climate resilient urban areas.

REFERENCES

Dal Bo Zanon, B., Roeffen, B., Czapiewska, K. M., De Graaf-Van Dinther, R. E., & Mooij, P. R. (2017). Potential of floating production for delta and coastal cities. *Journal of Cleaner Production, 151*, 10–20.

De Graaf, R. E. (2009). *Innovations in urban water management to reduce the vulnerability of cities.* PhD thesis, Technical University Delft, The Netherlands.

De Graaf, R. E. (2012). *Adaptive urban development: A symbiosis between cities on land and water in the 21st century.* Inaugural lecture. Rotterdam University of Applied Sciences, The Netherlands.

De Graaf-Van Dinther, R. (2019). *The four pillars of climate resilience.* Opinion Article. https://www.linkedin.com/pulse/four-pillars-climate-resilience-rutger-de-graaf-van-dinther/

De Graaf-van Dinther R (ed.) (2020) *Climate resilient urban areas. Governance, design and development in coastal delta cities.* Palgrave Pivot. Palgrave Macmillan.

De Lima R. L. P., De Graaf, R. E., Boogaard, F. C., Dionisio Pires, L. M., & Sazonov, V. (2015). *Monitoring the impacts of floating structures on the water quality and ecology using an underwater drone.* Conference Proceedings: 36th IAHR World Congress, 28 June–3 July 2015, The Hague, The Netherlands.

De Lima, R. L. P., Boogaard, F. C., & de Graaf-Van Dinther, R. E. (2020). Innovative water quality and ecology monitoring using underwater unmanned vehicles: Field applications, challenges and feedback from water managers. *Water, 12*(4), 1196. https://doi.org/10.3390/w12041196.

Ernst, L., De Graaf-Van Dinther, R. E., Peek, G. J., & Loorbach, D. A. (2016). Sustainable urban transformation and sustainability transitions; conceptual framework and case study. *Journal of Cleaner Production, 112*, 2988–2999.

Marris, E. (2017). Why fake islands might be a real boon for science. *Nature, 550*, 22–24.

Ovink, H. (2019). Personal communication. 22 Feb 2019.

Quirk, J. (2017). *Seasteading: How floating nations will restore the environment, enrich the poor, cure the sick, and liberate humanity from politicians.* New York: Free Press.

Roeffen, B., Dal Bo Zanon, B., Czapiewska, K. M., & De Graaf, R. E. (2013). *Reducing global land scarcity with floating urban development and food production.* Conference Proceedings: International Water Week, Amsterdam, 8 pp.

Trang, N. T. T. (2016). Architectural approaches to a sustainable community with floating housing units adapting to climate change and sea level rise in Vietnam. World Academy of Science, Engineering and Technology. *International Journal of Civil, Environmental, Structural, Construction and Architectural Engineering, 10*(2), 168–179.

Trent, J., Embaye, T., Buckwalter, P., Richardson, T.-M., Kagawa, H., Reinsch, S., & Martis, M. (2010). *Offshore membrane enclosures for growing algae (OMEGA): A system for biofuel production, wastewater treatment, and CO₂ sequestration*. NASA (Online Report) http://ntrs.nasa.gov/archive/nasa/casi.ntrs.nasa.gov/20100039342_2010039898.pdf

UN. (2019). *Sustainable floating cities can offer solutions to climate change threats facing urban areas, deputy secretary-general tells first high-level meeting*. Press release. https://www.un.org/press/en/2019/dsgsm1269.doc.htm. Visited 1 June 2020.

Von Wirth, T., Fuenfschilling, L., Frantzeskaki, N., & Coenen, L. (2019). Impacts of urban living labs on sustainability transitions: Mechanisms and strategies for systemic change through experimentation. *European Planning Studies, 27*(2), 229–257. https://doi.org/10.1080/09654313.2018.1504895.

Wang, C. M., Lim, S. H., & Zhi, Y. T. (2020). *Proceedings of the world conference on floating solutions. Lecture notes in civil engineering*. Singapore: Springer Nature.

Index[1]

---

[1] Note: Page numbers followed by 'n' refer to notes.

The manufacturer's authorised representative in the EU is Springer
Nature Customer Service Centre GmbH, Europaplatz 3, 69115 Heidelberg,
Germany. If you have any concerns regarding our products, please
contact ProductSafety@springernature.com

Printed and bound by CPI Group (UK) Ltd, Croydon, CR0 4YY
29/04/2026
02099471-0002